February 2013
(complementary)

Climate Change Politics

Climate Change Politics

Communication and Public Engagement

EDITED BY

Anabela Carvalho and Tarla Rai Peterson

CAMBRIA
PRESS

Amherst, New York

Copyright 2012 Cambria Press

All rights reserved
Printed in the United States of America

No part of this publication may be reproduced, stored in or introduced into a retrieval system, or transmitted, in any form, or by any means (electronic, mechanical, photocopying, recording, or otherwise), without the prior permission of the publisher.

Requests for permission should be directed to:
permissions@cambriapress.com, or mailed to:
Cambria Press
University Corporate Center, 100 Corporate Parkway, Suite 128
Amherst, NY 14226

Library of Congress Cataloging-in-Publication Data

Carvalho, Anabela.
Climate change politics: communication and public engagement / edited by Anabela Carvalho and Tarla Rai Peterson.
   p. cm.
Includes bibliographical references and index
ISBN 978-1-60497-823-0 (alk. paper)
1. Climatic changes--Political aspects. I. Carvalho, Anabela, editor of compilation.

QC981.8.C5C6168 2012
363.738'7456—dc23

2012032872

To Jarle, for unconditional love and support.
To Aspen, Gwen, and Zane, with hope for tomorrow.

Table of Contents

List of Tables ix

List of Figures xi

Acknowledgments xiii

Chapter 1: Reinventing the Political
        Anabela Carvalho and Tarla Rai Peterson 1

Part I: Public Engagement Through Social Marketing 29

Chapter 2: Climate Change Politics and Everyday Life
        Marianne Ryghaug and Robert Næss 31

Chapter 3: Landscape-based Discourse and Public Engagement
        Sarah Schweizer and Jessica Thompson 59

Chapter 4: The Visual Rhetoric of Climate Change Documentary
        Helen Hughes 87

**Part II: Public Engagement Through Public
Participation** 121

Chapter 5: State Commitment to Promoting Public Participation
        Anabela Carvalho and Joyeeta Gupta 123

Chapter 6: Climate Governance and Virtual Public Spheres
        Anna Maria Jönsson 163

Chapter 7: Web-based Public Participation
        *Andrea M. Feldpausch-Parker, Israel D. Parker, and Tarla
        Rai Peterson* 193

Part III: Public Engagement and Agonistic Politics 219

Chapter 8: Art and Political Contestation in Climate Issues
        Andrea Polli 221

Chapter 9: Visions of Climate Politics in Alternative Media
   Shane Gunster 247

Chapter 10: Poverty, Protest, and Popular Education
   Eurig Scandrett, Jim Crowther, and Callum McGregor 277

Chapter 11: Communicating for Sustainable Climate Policy
   Tarla Rai Peterson and Anabela Carvalho 307

References 321

Index 369

About the Contributors 375

List of Tables

Table 1: Open coding categories of communicating climate change at three national parks 69

Table 2: Latest available NCs for six countries 135

Table 3: Characteristics of the six countries analyzed. 138

Table 4: Three examples of climate change groups in Facebook, September 2010 173

Table 5: Political demographics 201

List of Figures

Figure 1: Going, Going, Gone (wayside interpretive sign at Glacier National Park) 82

Figure 2: HippyJimStarbrook and Greenpeace in Second Life 184

Figure 3: Greenpeace's Dove campaign in Second Life. 185

Figure 4: Human banner from the Stephansplatz rally in Vienna, Austria 203

Figure 5: Writing in the sand from the Yemen rally 211

Figure 6: Public Smog Offsets Tomorrow Today (Douala, Cameroon, 2009) 241

Acknowledgments

Some of the first ideas for this book emerged from conversations during Carvalho's sabbatical at the Institute for Environmental Policy Analysis (IVM) of the VU University Amsterdam, The Netherlands, in the spring of 2009. She would like to thank Fundação para a Ciência e a Tecnologia for her sabbatical stipend (SFRH/BSAB/901/2009). Conversations with Joyeeta Gupta, Dave Huitema, Eleftheria Vasileiadou, Frans van der Woerd, Constanze Haug, and others were especially stimulating. In the process of coediting a special issue for *Environmental Communication: A Journal of Nature and Culture*, Carvalho and Peterson discovered a mutual interest in pursuing a project that integrated analytical and aesthetic approaches to studying climate change communication as it contributes to the political (as defined by Mouffe). Peterson would like to thank Cristián Alarcón, Hans Peter Hansen, Nadarajah Sriskandarajah, and other members of the Environmental Communication Program at the Swedish Agricultural University for conversations probing possible relationships between communication and politics. She also would like to thank the other members of JET for keeping her on her toes when examining the interconnections among communication, policy, and politics related to climate change. Along the way, the following reviewers offered valuable suggestions for chapter revisions:

Finis Dunaway, Irene Lorenzoni, Eric Lee Morgan, Jennifer Peeples, Jean Retzinger, Chris Russil, Craig Trumbo. Several chapters in this book have profited from conversations associated with their presentations at the 2008 European Communication Conference (ECREA) in Barcelona, Spain; the 2009 Conference on Communication and the Environment in Portland, Maine; and the 2011 COCE in El Paso, Texas. We would like to thank all the authors for their patience with this very extended process. We appreciate their willingness to adapt individual styles to help create a stronger book. We are also grateful to Toni Tan of Cambria Press, who worked with us with astonishing grace and was always available to answer more questions.

CLIMATE CHANGE POLITICS

CHAPTER 1

REINVENTING THE POLITICAL

How Climate Change Can Breathe New Life into Contemporary Democracies

Anabela Carvalho and
Tarla Rai Peterson

Teetering on the edge

Scientific research increasingly indicates that climate change presents enormous threats to life as humankind knows it (e.g., Kiehl, 2011; Shakhova et al., 2010; Vermeer & Rahmstorf, 2009) but despite the economic slump of the last few years, the levels of greenhouse gas emissions continue to grow. Existing policies seem insignificant in the face of what is required to avert the worst impacts of climate change, and national and international politics appear to be plagued by perpetual impasse. While humanity stands on the edge of disaster, the main problem is a political one.

Climate change politics may involve more stakeholders than any other issue: Governments play a key part but so do international organizations, corporations, and nongovernmental organizations (NGOs). Moreover, climate change can be effectively addressed only through sustained citizen engagement. The far-reaching transformations that are needed to respond to climate change and to move to a different model of energy production and use require the involvement of citizens at the political level: decisions have to be made in a democratic way that is simultaneously inclusive and effective in the long term.

Despite politically influential actors' attempts to downplay its significance, society has found climate change impossible to ignore. The large increase in social-movement action on climate change in the last few years suggests that civic interest and participation are on the rise. At the same time, most people remain remarkably disengaged from climate change politics. *Climate Change Politics: Communication and Political Engagement* is concerned with the ways climate change communication may contribute to a transformation of politics. It examines connections between a sense of political powerlessness and the symbolic environment in which current democratic politics is enacted, as well as how a rapidly changing communication context may open new political possibilities as it closes off others. With widespread transformations in the processes of production and consumption of mediated messages, new spaces for political interaction are emerging.

Climate change communication has the potential to induce significant transformations in civic politics. As issues that cut across so many sectors and scales, climate change and communication about climate change are redefining the boundaries between public and private, political and domestic, subject and community. Just as political action may affect climate change, climate change may affect the political. This book explores how people's sense of engagement with collective problems may change as they address the questions posed by climate change, how new political identities for citizens develop in response to the dilemmas involved in climate change, and how the meaning of climate change is redefined in communicative practices in ways that sediment or instead shake the edifice of politics and the relations between politicians and citizens.

This exploration of connections between communication and the political includes analyses of how people represent, construct, and circulate ideas about climate change and of how these practices translate into decisions and public policies, as well as how they relate to political identities. Given the multifaceted nature of climate change, multi- and interdisciplinary approaches are needed to understand different dimensions of the issue and their interconnections. Drawing on a variety of research fields, from media and political sociology to popular education, this book initiates a dialogue between theoretical traditions that enrich the comprehension of relations between communication and the politics of climate change.

The chapters in this book examine various forms of climate change communication: in artistic expression ranging from installations to cinema, in web-based spaces and in alternative—that is, nonmainstream—media. They analyze a range of semiotic resources and practices within which the meanings of climate change are negotiated. All chapters examine social transformation as it links with communication practices, or how issues are managed and by whom based on certain hegemonic discourses. They describe ways some discourses become hegemonic and others are marginalized and how this shifting symbolic landscape translates into political subjectivity. By looking at the multiple ways people experience and represent climate change, we take the analysis beyond the cognitive to include emotional, aesthetic, and other epistemologies that shape political engagement with this issue.

In this chapter we first explore how the ironic position of politics in a society that has been labeled *postpolitical* constrains climate change politics. We then suggest how citizen engagement may reinvigorate the democratic potential of politics by redirecting attention. Finally, we offer a preview of the cases analyzed here, which illustrate possible routes for this redirection.

The political in climate change politics

In a special issue of *Theory, Culture & Society* published in 2010 in which a set of prominent thinkers reflected on climate change, Erik Swyngedouw pointed out an apparent paradox: Though climate change and its consequences have been elevated to key policy issues in the last few years, many analysts claim that social life has become depoliticized in contemporary democracies, that dispute and confrontation have been erased from the public sphere and replaced by technocratic and consensually framed modes of management of public problems. As Swyngedouw noted, scholars such as Chantal Mouffe, Jacques Rancière, and Slavoj Žižek have offered critiques of this situation, which they have termed postdemocratic or postpolitical (Mouffe, 2005; Rancière, 2006; Žižek, 1999).

So what kind of politics is associated with climate change? Swyngedouw argued that through its presentation in apocalyptic terms and its reduction to a problem of CO_2 emissions, climate change itself has given rise to a hegemonic populist proposal that promises solutions within the structures of capitalism and the market economy. The framework of "sustainability" offers what appears to be the only way out and does so without requiring any fundamental social or political change. Swyngedouw maintained that the hegemonic discourse of sustainability threatens to overdetermine the political options:

> the sustainability argument has evacuated the politics of the possible, the radical contestation of alternative future socio-environmental possibilities and socio-natural arrangements, and has silenced the antagonisms and conflicts that are constitutive of our socio-natural orders by externalizing conflict.
> (Swyngedouw, 2010, p. 228)

Others have previously suggested that the discourse of sustainable development has become hegemonic because it is a consensual language that gets its force—in part, at least—from its ambiguity (Luke, 1995;

Peterson, 1997). It has a disciplinary role in relation to more radical forms of environmental discourse and mobilization, annihilating the possibility of opposition. Several studies have demonstrated that the related discourses of sustainable development and ecological modernization have been naturalized and neutralized by the media, a key element of contemporary public spheres (Carvalho, 2005b; Carvalho & Pereira, 2008), excluding more socially transformative discourses.

Swyngedouw maintained that a similarly hegemonic approach to climate change has contributed to the already ongoing depoliticization of public life. Dissent and disagreement have no space in a managerial framework that claims to "solve the problem" while leaving comfortable lifestyles untouched. Instead, a series of mechanisms have been put in place to regulate and trade emissions of greenhouse gases through institutional arrangements and voluntary measures. This can be viewed as a form of governmentality (Dean, 1999; Foucault, 1991; Rose, 1999) that spans multiple scales and involves different agents through techniques of domination and technologies of subjectification.

Looking at a range of texts from international organizations, Methmann (2010) has argued that climate protection has been transformed into an empty signifier (a concept employed by Laclau & Mouffe, 1985) that is employed by a multitude of bodies to justify their continuation of practices that have, paradoxically, contributed to the problem. Thus the World Trade Organization (WTO), for instance, frames climate change as a trade problem and promotes further liberalization to address it. The International Monetary Fund (IMF) promotes a green fund as a technical solution to the problem that it argues would be more effective than political negotiations. Both bodies, together with the Organization for Economic Cooperation and Development (OECD) and the World Bank, construct economic growth as the priority in any climate-oriented policies, which are therefore expected to promote growth. Climate protection is thus integrated into "the global hegemonic order without changing the basic social structures of the world economy" (Methmann,

2010, p. 348). "Put bluntly, climate mainstreaming fits well in the overall project of sustainable development, that is, 'sustaining capitalism' in its present condition" (p. 369).

These accounts suggest that despite the unique nature and scale of climate change, the primary responses have been limited to politics as usual and have deepened dominant modes of governing the world. Is this all there is in climate change politics? Where is the political in climate politics? Inspired by the writings of Chantal Mouffe (e.g., 1993/2005, 2000, 2005), we describe *the political* as engagement with processes of debate and decision making on collective issues in which different values, preferences, and ideals are played out and opposed.[1] Given this definition, climate change communication as described by Swyngedouw (2010), Methmann (2010), and others (Carvalho, 2005a; Paterson & Stripple, 2010) may indeed have evacuated all political dimensions from this ironically political issue.

As much as Swyngedouw's (2010) and Methmann's (2010) perspectives do justice to the way many things are run, there are contrasting tendencies in the wider politics of climate change. In fact, activism on climate change has emerged as one of the most prominent forms of citizen-led politics (Endres, Sprain, & Peterson, 2009). In the months and years of the run up to the United Nations Framework Convention on Climate Change (UNFCCC) 15th Conference of the Parties (COP 15) in December 2009, a large social movement developed around the goal of achieving an ambitious international agreement to curb greenhouse gas emissions. It converged in Copenhagen in the form of tens of thousands of demonstrators, a display of public mobilization of historic proportions. The achievements of campaigns such as 350.org (see chapter 6) and Tck tck tck, the World is Ready in organizing hundreds of events all around the world before COP15 are also worth mentioning. The resounding failure of the summit to shape new international policies certainly dampened some of the civic energy around climate change. But civic action extends far beyond occasional street demonstrations. From protest groups such as Plane Stupid to community-based projects such as Transition, a range of social initiatives has emerged and expanded significantly, both within single countries and internationally. There is also a growing public involvement with what has been termed "click-activism," Internet-mediated forms of political expression and participation. Despite these developments, however, polls show that most people feel powerless to address climate change and feel disengaged from climate politics (Yale Project on Climate Change & the George Mason University Center for Climate Change Communication, 2009). The locus of climate politics is perceived as removed from individuals' sphere of influence, and the global scale of the problem deters social mobilization and action. Discursive practices, especially those that dominate mass media, may play a role in the development of this form of political subjectivity (Carvalho, 2010).

This book offers a critical yet hopeful examination of opposing signs of political vitality in the politics of climate change and discusses how people use various forms of critique, contestation, and alternative thinking to challenge the hegemonic technomanagerial approach. Because the meanings of climate change and of the numerous aspects of reality associated with it are constructed through communication, we offer an analysis of communication practices and structures as constitutive of climate change politics. The options considered and the choices made result from social interaction based on communication, whereby people play out different values and worldviews. The roles and identities of participants in those processes, such as the state and citizens, are also defined and redefined in communication.

COMMUNICATION AND PUBLIC ENGAGEMENT IN THE POLITICS OF CLIMATE CHANGE

Given the nature and the scale of climate change, there is a widespread acknowledgment that citizen engagement is indispensable to finding effective responses. Nonetheless, as illustrated by Whitmarsh, O'Neill, and Lorenzoni's (2010, pp. 2–3) recent book, understandings of citizen engagement vary significantly, as do the goals and values of those *engaged in engaging* the public:

> Most Western governments have an interest in engaging the public in debate about the type of society they want to live in and empowering communities to bring about change to that effect. Here, the focus is on public participation in policy making, community decision making and grassroots innovation.... For other groups there may be different reasons for being interested in the public's understanding of and responses to climate change. Businesses may be involved with formal climate change communication as part of a corporate social responsibility or a product marketing agenda (and often both); and nongovernmental organizations (NGOs) may do so because climate change intersects with their existing environmental and social concerns and interests.

We argue that at least three modes of public engagement are associated with different views of climate change politics and communication. They range from *social marketing* and *public participation* to *agonistic pluralism*. The first mode of public engagement with climate change, which provides the lightest form of social intervention, involves persuading individual citizens to voluntarily modify some aspect of their energy-related behavior or accept some policy proposal that typically stays within the limits of existing economic and institutional structures (Moser & Dilling, 2007; Nisbet, 2009). In the last decade or so, multiple social actors have promoted the idea that "we should all do our bit." From government to civil groups, numerous organizations call for small acts, such as switching to low-consumption lightbulbs and unplugging appliances not in use. Some throw flying in the spotlight and advocate carbon offsetting. In all these examples, responsibility for climate change is individualized and the political realm is reduced to lifestyle choices. People are typically addressed as individual consumers rather than as communities that can act collectively on climate change.

Attempts to engage the public in this first mode typically involve social-marketing tools. Market research, cognitive theory, and strategic communication are employed to appeal to individual behavior change. For example, Greenpeace has used the polar bear to inspire people to act on climate change (Slocum, 2004). A range of other organizations, from the United Nations Environment Program (UNEP & Futerra, 2005) to WWF-UK, the Climate Outreach and Information Network, the Campaign to Protect Rural England, Friends of the Earth, and Oxfam (Crompton, 2010) have attempted to ascertain the optimal communication strategies for environmental matters. O'Neill and Nicholson-Cole (2009) explored the utility of fear as a motivator for individual action to mitigate climate change, and Nisbet and Kotcher (2009) documented the use of personalized blogs to develop and maintain a strong relationship with potential climate change activists.

Social-marketing campaigns emphasize communication's instrumental, rather than its constitutive, dimensions and serve the purpose of persuasion by employing demonstrably effective communication formulas in a top-down process. Consistent with this perspective, Lakoff (2010) and Nisbet (2009) claimed that appropriate framing is a basic tool for persuading citizens to support climate change policy. Moser and Dilling (2007) asserted that motivating individual action through persuasive communication and education offer a powerful basis for social change needed in response to anthropogenic climate change.

Brulle (2010), however, maintained that by professionalizing communication and thereby turning communication experts into the arbiters of environmentalism, social marketing further excludes citizens from the processes of decision making that are needed to deal with climate change (see also Corner & Randall, 2011). Similarly, Paterson and Stripple (2010) argued that social marketing can be read as a form of governmentality (cf. Dean, 1999; Rose, 1999). Brulle (2010) argued that social-marketing campaigns fail to promote the political mobilization that would enact the vast transformations necessary to address climate change. In his view, by focusing on "short-term pragmatic actions that fit within economic and political imperatives" (p. 83), these strategies actually undermine efforts to address climate change. This mode of public engagement functions well within the logic of economically liberal democracies. It fits the hegemonic technomanagerial discourses of sustainable development and ecological modernization that we discussed earlier.

The second mode of public engagement involves enhancing the role of citizens in policy-making processes through public-participation arrangements. In the last few decades, multiple voices in academia and in the political sphere have called for forms of governance that further the participation of the public in political decision making. The movement for Public Engagement with Science and Technology has promoted the inclusion and the participation of citizens in matters that involve scientific and technologic knowledge and in corresponding political decisions (e.g., House of Lords Select Committee on Science and Technology, 2000). Citizen participation in environmental issues has been promoted as the guarantor of better and more accountable decisions (e.g., Coenen, 2010; Dietz & Stern, 2008). In political arenas, there is increasing recognition that the strength of democracies and the quality of decisions is a function of public participation and engagement. The U.S. National Environmental Policy Act (1970), the Rio Declaration on Environment and Development (1992), and the Aarhus Convention on Access to Information, Public Participation in Decision-making, and Access to Justice in Environmental Matters (1998) have all decreed the importance of public participation and access to information on environmental issues.

As Arnstein pointed out in her seminal 1969 article, the "ladder of participation" has several steps, and public participation is arranged in different shades, from information and consultation to deliberation. Although participation has mainly taken the form of information gathering and public consultation, various countries have experimented with citizen juries, consensus conferences, and other deliberative formats of public participation (e.g., Kleinman, Delborne, & Anderson, 2011).

However, the impacts of these exercises are limited. The institutional and sociocultural conditions for dialogical communication between citizens, researchers, policy makers, and other stakeholders have rarely been fulfilled, and public participation is often viewed as a mere legitimation tool that can be quite exclusionary (e.g., Besley, 2010; Braun & Schultz, 2010).

Delgado, Kjølberg, and Wickson (2010) have argued that there are tensions and ambiguities regarding rationales of public participation, who is a relevant participant, pros and cons of invited versus self-organizing citizen groups, the stage at which citizens should be involved, and context specificity versus transferability of public-participation models. In a study of discourses on civic engagement with climate change in Denmark, Lassen, Horsbøl, Bonnen, and Pedersen (2011) confirmed these problems; they found vagueness and a lack of clarity regarding citizen roles in policy processes. One case they studied was led by a Danish municipality, Frederikshavn, that aims to rely exclusively on renewable sources of energy by 2015 and that has called for citizen activism on climate change to help reach that goal—a paradoxical appeal, given that activism is normally a means of protesting, rather than buttressing, centralized control.

Another initiative worth mentioning that attempted to involve citizen voices at a different scale is World Wide Views (WWViews, n.d.), a "global" consultation process led by the Danish Board of Technology. Some 38 countries took part in organizing local deliberations on September 26, 2009. The process involved group discussions on various aspects of climate change and the formulation of a consensus message for negotiators at COP15. The stated goal of WWViews was "to give a broad sample of citizens from across the world the opportunity to influence the COP15 negotiations and thereby the future of global climate policy ... [and] to demonstrate that political decision-making processes on a global scale can benefit from the participation of ordinary people" (WWViews, n.d.). Analyzing the Danish local deliberation event, Phillips (2011) has shown that the dynamics of inclusion of citizens and the inevitable exclusion of many voices constrain the practice of deliberation in top-down approaches to the production of consensus.

Public participation in policy processes, in the case of WWViews, has been predominantly advocated in the context of models of deliberative politics (e.g., Dryzek, 2000; Gastil & Levine, 2005; Habermas, 1996). Drawing on Habermasian notions of communication, these models assume that it is possible and desirable to achieve consensus on collective issues through rational argumentation.[2] Also designated *discursive politics*, this view places faith in the possibility of reaching agreement through argumentation. Communication is here viewed as mainly a process of cognitive influence through information and good reasoning. Chantal Mouffe (e.g., 1993/2005) has challenged this view of democratic politics, maintaining that rational decisions cannot be reached through communicative exchanges and that a fully inclusive consensus is not possible. As power differences always play in political relations, any consensus is exclusionary, and therefore it is an illusion that harmonious solutions that satisfy all can ever be reached. Lassen et al. (2011) and Phillips (2011) provided specific illustrations of these tensions in climate politics.

Public participation in politics is not necessarily limited to formal political arrangements in which people's roles and the possibilities of intervention are predefined. It also can take the shape of uninvited action, such as media campaigns, public debates, or demonstrations in which "self-selected actors ... turn into participants through collective actions" (Braun & Schultz, 2010, p. 407).

Whereas social marketing and formal public participation are top-down managerial practices, citizen-led political participation is initiated from the bottom up. Engagement starts with citizens who see faults in the ways formal political institutions deal with climate change and advance alternative forms of governance, whether through proposals for different governmental policies or through social and economic changes.

This involves dissent over alternative political projects. The third mode of engagement cultivates political conflict and rejects the viability of consensus between opposing viewpoints.

Peterson, Peterson, & Peterson (2005) noted that officially sanctioned public-participation practices often require bracketing conflict and thus involve treating the current political hegemony as truth. As Mouffe (2000) contended, the illusion of consensus is fatal to democracy because a healthy democratic process requires recognition of differing interests and the recognition that open conflict over differing interests is legitimate. Turning to the relations between discourse and politics, Ivie (2004, p. 21) specified that "democratic dissent ... is as alarming to the purveyors of prevailing opinion as it is critical to a nation's political welfare." Ivie (2004) and Peterson et al. (2005) added that dominant elites generally prefer consensus-based approaches to those based on debate because they have access to sufficient resources to hold out until some semblance of consensus is reached. Reliance on squelching argumentation not only jeopardizes democracy by legitimizing existing hegemonic configurations of power and precluding resistance against dominant elites, however; it also artificially reduces power relationships to superficial conflicts of interest, presumably reconcilable through mutual good will. Without debate regarding their political dimensions, for example, government practices that contribute to anthropogenic climate change are naturalized. In the absence of such debate, existing hierarchies become uncontested reality rather than outgrowths of neoliberal economic goals with serious implications for the global climate.

The notion of politics that underpins the third mode of engagement with climate change places difference and confrontation at the center of public life. Chantal Mouffe (e.g., 1993/2005) argued that conflict and antagonism are inevitable in social life and that it is necessary to envisage ways to create a pluralistic democratic order that can make space for those differences.

> Liberal democracy requires consensus on the rules of the game, but it also calls for the constitution of collective identities around clearly differentiated positions and the possibility of choosing between real alternatives. This agonistic pluralism is constitutive of modern democracy, and rather than seeing it as a threat, we should realize that it represents the very condition of existence of such democracy.
>
> (Mouffe, 1993/2005, p. 4)

The notion of agonistic politics builds on a distinction between antagonism (between enemies) and agonism (between adversaries). Whereas an enemy is to be excluded from the polis, an adversary's existence is accepted as legitimate, as is the adversary's right to defend his or her ideas.

Advancing the project of a "radical and plural democracy," Mouffe (1993/2005, 2000, 2005; see also Laclau & Mouffe, 1985) advocated conceptualizing democracy in terms that not only allow for but also encourage confrontation and pluralism. She offered the democratic paradox as an alternative to the universalist-rationalist perspective (Habermas, 1996; Rawls, 1993) that has dominated political theory since the last half of the 20th century. Whereas the universalist-rationalist perspective shares with Enlightenment science the aim of establishing universal truths independent of context, the agonistic perspective contends that the "illusion of consensus and unanimity" is fatal to democracy because a "healthy democratic process" requires the "vibrant clash of political positions and an open conflict of interest" (Mouffe, 2000, p. 130). Agonistic politics entails continual negotiation of the democratic paradox between individual freedom and community equality, without seeking an impossible reconciliation between the two. It supports neither accepting all views as equally valid, which leads to anarchy, nor building an artificial consensus, which leads to authoritarianism and tyranny. To avoid both anarchy and tyranny, a polity must find ways to enable political clash, and it must do so in a way that ensures appropriate representation of community members who are less likely to be heard (Peterson, Peterson, & Peterson, 2005; Peterson, Peterson, Peterson, Allison, & Gore, 2006). As Ivie (2001) demonstrated, dissonance is needed to invite contestability and to critique the terms that threaten to compromise political difference.

An agonistic approach to democratic politics sets aside the assumption that once its citizens become rational every nation will eventually accept a similar brand of liberal democracy. Rather than constructing democracy as a fragile ornament that must be protected, curtailed, or even suspended when chaos threatens, that must be deferred or delegated until divisive circumstances subside and a political culture made up of reliably informed and rational citizens finally emerges, it proposes democracy as an open clash of political positions that may occur today (Ivie, 2004, 2011).

Although centrist approaches, such as the Washington Consensus in the United States and the Third Way in the European Union, enjoy political success, the consensus-based democracy they represent weakens both liberty and freedom (Mouffe, 2000). The focus on meeting in the middle has meant giving up both the quest for equality and the quest for liberty and has been achieved only by relegating emotional and aesthetic epistemologies to an extrapolitical sphere. In such a context, it makes perfect sense to proclaim a postpolitical world, for a value-free politics is irrelevant.

Agonistic politics holds special potential in the case of climate change communication because climate change is the opposite of the ideal situation for consensus. Consensus processes are ideal for situations in which "scientific information about an environmental issue has high predictive power and its application is relatively uncontested" (Peterson, Peterson, & Peterson, 2005, p. 766). In disputes that are laden with sharp power differentials and complex uncertainties, however, the drive toward consensus reinforces public apathy and cynicism, and reinforces existing power relationships (Peterson, Peterson, & Peterson,

2006; Toker, 2005). Given that climate change policy is the antithesis of an issue that is likely to be resolved by consensus decision making, citizen dissent emerges as extraordinarily important.

Political struggles, be they more or less apparent, are inscribed in a diversity of communication practices, from political speeches to mainstream and alternative media and artistic forms. As suggested by numerous scholars (e.g., Fairclough, 2003; Ivie, 2001; Wodak, 2011)—whose understandings of discourse do not necessarily coincide—discursive processes can contribute to solidifying the political status quo but can also help transform it. On the one hand, certain discursive constructions sediment, harden, and become dominant, appearing natural and objective. They constrain the range of legitimate political options, what is viewed as acceptable, doable, and thinkable at a given moment in a given society. On the other hand, it is (almost) always possible to challenge and modify political ideas and political arrangements through alternative discourses that may be created and promoted by certain individuals, groups, or critical communities. Political and institutional change takes place when hegemonic discourses are delegitimated and alternative ones gain social acceptance.

In agonistic politics, discourse is seen as fundamentally constitutive of the social. In fact, Laclau and Mouffe's (1985) notion of discourse is not limited to language but encompasses all social phenomena. In their antiessentialist view of the social world, the meaning of objects and actions is constructed through changing systems of difference. Partial (and temporary) fixation of meaning occurs through "articulatory practices" that involve the construction of "nodal points," but it allows different discourses to emerge that may challenge and modify understandings of the world. This poststructuralist approach leaves space for agency. Social subjects are inscribed in a diversity of communication practices that create different subject positions. However, subjects have room to question and modify dominant discourses, including their role therein. Because individual and group identities, including one's condition as a political subject, result from discursive processes and structures, they are contingent and not fully fixed. Hence political subjectivity depends on subject positions in a discursive structure, but possibilities for action remain always within that structure.

We maintain that combining the study of politics with communication is vitally important to understanding social processes, especially around complex issues, such as climate change. Integrating discourse theory, as well as other contributions from communication studies, into the analysis of political matters can make a positive contribution in that direction. As this volume illustrates, different ways of viewing communication, with specific theoretical and epistemological underpinnings, generate diverse readings of climate change politics that complement each other.

BOOK PREVIEW

Chapters in this book account for all three modes of engagement with the politics of climate change. Some cases primarily discuss communication strategies played out within the limits of top-down agendas, but others refer to alternative understandings of climate change politics that have been developed or advanced by citizen groups. Categorizing the chapters that follow comes with all the attendant weaknesses of any modernist system for describing reality. It oversimplifies relationships among the essays, creating false dichotomies between those that could fall into multiple categories, and it masks differences among those we place in the same category. We are certain there is room for argument regarding whether any individual essay is best described as representing one category as opposed to another. The point is not to use a Procrustean approach to force essays into particular categories so much as to find a way of organizing the analyses that helps the reader think about how they engage each other and with our larger concern about the politics of climate change communication. Chapters 2, 3, and

4 focus primarily on engagement through social marketing, and chapters 5, 6, and 7 look at engagement mainly—albeit in no way exclusively—as public participation. Chapters 8, 9, and 10 focus on engagement through agonistic pluralism, whereas chapters 4 through 10 all suggest connections between either social marketing or public participation and agonistic pluralism. Further, this organizational framework highlights a question suggested earlier. At least since the late 1980s, climate change has assumed an increasingly central and contested role within public discourse. The essays in this volume examine how climate change communication has shaped and been shaped by the contemporary political landscape. Our intent is to demonstrate ways that each of these forms of public engagement may contribute differently, yet complementarily, to a democratically informed strategy for bringing politics to bear on the issue of climate change.

This book explores cases from a wide range of countries, including Norway, Sweden, the United States, China, India, and Yemen. These countries are in very different stages of economic development, represent diverse forms of political governance, and also have different legal-political statuses in the international regime on climate change: In the context of the Kyoto Protocol to the UNFCCC (which the United States did not ratify), the first three are expected to reduce their greenhouse gas emissions whereas the latter three are considered developing countries that are therefore exempt from that commitment. The multiscalar nature of climate change is often described as a challenge to effective mitigation politics. Some chapters in this book focus on the national scale but several cut across geographical scales, looking at the interplay between various local, national, and global spaces of communication and political action.

Chapter 2 offers a critical diagnosis of the possibilities of engagement with climate change from the point of view of "ordinary citizens"—that is, people who are not actively involved in social or political movements to address the problem. Ryghaug and Næss discuss the ways in which Norwegian citizens relate to top-down communication on climate change and how they construct their roles and responsibilities, as well as roles and responsibilities of politicians, in mitigating the problem. As Laclau and Mouffe (e.g., 1985) argued, meanings are never fully fixed: They are dependent on specific uses of language (comprising verbal, visual, and other semiotic resources), as well as on social and cultural contexts; similarly, identities are relational, dependent on communication practices and therefore unstable. Ryghaug and Næss examine the ways in which climate change and people's perceived powerlessness to address it are discursively organized and (re)negotiated in the domestic context of people's daily lives. They show that understandings of climate change and of climate change politics are strongly linked to media discourses, to political discourses, and to people's living conditions, and that political critique and disengagement result from perceived contradictions in political management of and (in)action regarding the problem.

Schweizer and Thompson offer important insight into place-based public understandings of climate change. Among its other challenges, engaging climate change is complicated by perceptual matters, such as disconnection from the natural world, which prevents many people from noticing subtle changes in the environment (Moser, 2010). By looking at (the potential for) communication about climate change in U.S. national parks, they address the geographical challenges that were pointed out earlier and make crucial suggestions in terms of both cognitive and affective dimensions of engagement with climate change (cf. Lorenzoni, Nicholson-Cole, & Whitmarsh, 2007), which can be politically mobilized or activated at the individual level. They note that citizens are exposed to many messages about climate change on a daily basis. Research has shown that these messages resonate more effectively when they are meaningful to the audience and framed consistently with cultural values and beliefs, and when they suggest specific actions that audience members feel comfortable taking. Based on place-attachment theory and experiences with place-based education, the authors theorize that landscape-based discourse about climate change in U.S. national parks has the potential to create new and productive space for political action on climate change. Likewise, they use insight from place-based education and experiential learning to explain the rhetorical power of national parks in the United States. Given their compelling cultural presence in the shared (both virtual and material) experience of self-identifying as an "American," Schweizer and Thompson suggest that climate change communication in U.S. national parks has the potential to promote public awareness and to suggest lifestyle modifications that may help mitigate climate change impacts. Their analysis of ways that national parks in the western United States currently communicate about climate change leads them to conclude that although communicators in some parks use the appeal of the material place to create a site for learning about climate change impacts and adaptations by merging visitors' personal experiences with national identity, communicators in other parks completely miss the opportunity. Though Schweizer and Thompson's analysis emphasizes engagement through social marketing and therefore focuses on communicative instrumentality, their study also showcases the largely unrealized communicative potential for constituting a particular national identity from individual visitor experiences. Ultimately, they suggest ways a physical, federally managed space can function as a forum for public engagement in climate change policy by contributing to a deeper understanding of ecological impacts and potential management responses.

Hughes examines the role of aerial cinematography in contemporary climate change documentaries. She marries analysis of a particular shot type, which tends toward politicizing aesthetic issues, with an understanding of how it may contribute to a strategy for communicating about climate change. Her discussion of long shots in cinematography shows how media can reframe images of human impact on the world and explores implications of this constitutive potential for engagement with climate change. She notes that contemporary documentary filmmakers use high-angle extreme long shots, including aerial shots and space photography, to present evidence for the considerable effects of centuries of agriculture and industry on the environment. The visibility of large-scale landscape interventions and atmospheric effects generates spectacular visual content that can be used to persuade audiences of the material reality of climate change. These vast images offer one way to solve the conundrum of how to signify sufficient immediacy to achieve personal salience without negating awareness of the suprahuman temporal and spatial scales of climate change. Enabling people to comprehend a complex global phenomenon that is plagued by uncertainty may be cause for celebration, and yet two quite distinct and opposing directions can be discerned in 21st-century attitudes toward aerial images. The first takes the optimistic view that the rhetoric of environmentalism, supported by still and moving images—especially of the whole planet—is creating a growing activist response from audiences. The second is a more pessimistic concern that still and moving image technologies, integral to the predominantly visual culture of modernity and particularly significant in the development of remote-control surveillance and weaponry, themselves contribute to the distortion of habitable space. Hughes analyzes the perspective found in the use of aerial photography, aerial cinematography, space photography, and satellite images in three U.S. climate change documentaries: *An Inconvenient Truth* (David & Guggenheim), *Everything's Cool* (Gold & Helfand), and *The 11th Hour* (Conners, Petersen, & Conners). She points out that these documentaries use extreme long shots to materialize otherwise abstract concepts drawn from climate change research. Environmental documentary film producers have recognized the potential of cinematic long shots not only to persuade viewers of the urgency of the situation regarding climate change but also to present possibilities for responding individually and collectively to it.

In chapter five, Carvalho and Gupta turn readers' attention to the explicitly political arena with their analysis of policy documents from signatories to the UNFCCC. Building upon Schweizer and Thompson's claim that public engagement is a crucial (but not necessarily a sufficient) motivator for developing national and international policy to mitigate anthropogenic climate change, Carvalho and Gupta argue that citizen engagement is a key factor in identifying what possibilities for mitigation and adaptation to climate change are acceptable for national governments to pursue. This engagement delineates the scope for both individual and political decision making, as well as for the successful implementation of any measures. Although challenging, public participation in policy processes has been shown to improve the quality of political decisions through the inclusion of alternative problem definitions and forms of knowledge. Recognizing that political efforts toward climate change policy are futile without strong public engagement, the UNFCCC has committed its parties to promote and facilitate "the development and implementation of ... public participation in addressing climate change and its effects and developing adequate responses" (article 6). One of the instruments to assess the implementation of such a commitment is the national communications reports to the UNFCCC. Carvalho and Gupta present a comparative analysis of the national communications of six countries: China, India, Portugal, Tuvalu, the United Kingdom, and the United States. Those countries make very different contributions to global greenhouse gas emissions and have different degrees of vulnerability to climate change, making them significant case studies. The analysis of national communication reports examines what countries have done to promote public participation in climate change action and how the reports discursively construct citizens' status and identities, as well as state–society relations.

The two following chapters share a focus on how various groups have used the Internet in response to the challenging spatial scale of climate change communication. First, Jönsson (chapter 6) and Feldpausch-Parker, Parker, and Peterson (chapter 7) account for forms of Web-mediated activism that involve some form of collective organization and action suggesting that communication structures may significantly enhance possibilities of engagement with the politics of climate change. Second, both chapters analyze forms of engagement with climate change that span several spatial scales and that in some ways can be considered global, a political space that can be especially difficult for citizens to access.

Jönsson's study of Facebook and Second Life shows that citizens, both individually and in the context of NGOs, are developing alternative forms of governance in virtual spaces. These online communities can be viewed as bottom-up political spaces. Though they have little impact on politics "in real life" (IRL), these developments will likely influence political subjectivities and are therefore significant. New information and communication technologies, and especially the Internet, offer an important potential to remake the political. In fact, many have associated the Internet with a revival of democracy by highlighting its role in the mobilization of social movements and political activism (e.g., Atton, 2004; Dahlberg & Siapera, 2007). Yet we remain skeptical about the claim that the Internet alone will bring the world closer to Mouffe's radical and plural democracy. Although it represents a potentially more open and more democratic alternative to traditional communication opportunities, the Internet also creates new divides along age, income, and country lines. This poses new questions. Whereas Jönsson convincingly makes the case for transnational deliberative politics, we evoke Mouffe's critique and note that exclusionary processes are likely to be even more acute at this scale and consensuses are likely to be even more fraught with distortions. Still, as Jönsson points out, online spaces can enhance political expression beyond the constraints of national boundaries and can contribute to the development of new forms of civic debate.

Feldpausch-Parker, Parker, and Peterson examine 350.org, a campaign promoting a limit of 350 parts per million of CO_2 equivalent in the atmosphere, largely below the 450 ppm limit that is roughly associated with the 2 °C target proposed by the European Union and other players that are considered to be at the political forefront of climate policy.[3] In itself the 350 parts per million goal is a radically different political target that would require substantive transformations at the levels of government, economy, and society. Moreover, the campaign involved a large number of (communicative) events that were conceived and put in place by citizen groups as politically meaningful acts. Feldpausch-Parker et al. draw on Giddens's structuration theory to argue that the public can mobilize resources, including but not limited to communication, to influence policy makers. Apart from the potential for actual transformation of (inter)governmental politics, this project suggests that citizens around the world feel politically motivated to self-constitute into agents of protest and debate on climate change, challenging the claims that current democracies are postpolitical. The 350.org initiatives aimed to offer an alternative to governmental politics. Although they are very diverse, all those communicative actions render the citizen an actor of the political—that is, the confrontation of proposals to respond to collective problems.

In the following chapter, Polli explores another form of political agency: art as communication on climate change that can promote awareness of, critiques of, and alternative options to hegemonic positions. Her essay discusses the implications of different models to manage access to, use of, and—as she argues—ownership of the air. She takes issue with technomanagerial constructs, such as emissions markets and trading and the privatization of the atmosphere. The chapter points out that art may have contributed to the cultural acceptance of the commodification of air and other immaterial resources, and she then moves on to discuss the work of a number of artists who have created alternative visualizations of problems associated to air pollution and fossil fuels. They unsettle understandings of causation, responsibility, and commercialization; some artworks specifically engage emissions trading and interactively expose its flaws. As Mouffe (2007) has argued, art can play a critical role in undermining the program of total social mobilization of capitalism, which permeates climate politics at multiple levels. Polli shows that politically dominant discourses on the atmosphere can be questioned and subverted, and that art can be focal to oppositional struggles and to remaking the political in climate politics. Most artworks referenced in the chapter encourage public participation in the politics of the atmosphere and of climate change, thereby advancing democratic pluralism. Fomenting dissensus and giving voice to those who are silenced by the existing hegemony (Mouffe, 2007), critical artistic practices can help foreground the political—rather than the technical—nature of responses to climate change and thus reinvigorate public engagement.

Gunster's chapter turns attention to alternative media spaces and their specific ways of framing the politics of climate change. News media have been shown to function as the public's main source of information on climate change (e.g., Cabecinhas, Lázaro, & Carvalho, 2008; Wilson, 1995). The media produce, reproduce, and disseminate multiple discourses on climate change in which knowledge, values, and power issues come into play, and they occupy a central position in the public space of contemporary societies, a battleground where different hegemonic projects are confronted (Mouffe, 2007). Their contribution to the social construction of politics—as well as of the political—warrant detailed analysis. Although a number of studies have focused on the coverage of climate change in mainstream media, alternative media are blatantly underresearched. Looking at two independent Canadian newspapers, Gunster's analysis explicitly foregrounds the political dimensions of climate change. Whereas news in corporate media tends to be dominated by narratives of political failure to tackle climate change, Gunster suggests that alternative media often offer more optimistic, even if critical, images and argues that media accounts of developing successful governmental projects for the generation of renewable energy or for the reduction of greenhouse gas are likely to stimulate the perception that all governments are capable of undertaking effective action; those accounts thus promote civic pressure for that to happen. He finds a strong emphasis on political rather than technological or lifestyle-related solutions in alternative media, which often involved contrasting actual politics with the politics of the possible, that is, referring to beneficial measures that *could* be put in place but are being ignored by governments. Another unique characteristic of alternative media when compared to mainstream media is their readiness to challenge dominant paradigms of economic growth. Gunster notes that, unlike their corporate counterparts, alternative media often advance radically different ways of thinking about the economy, culture, and society so that effective responses to climate change can emerge. Finally, and most important, this chapter explores the role of specific practices of mediation of the political for citizen-led politics. In the mainstream media, politics tends to be the exclusive domain of national and international elites, whom citizens can only passively observe rather than influence. In contrast, independent media construct political activism as effective and desirable. By offering in-depth stories of demonstrations, sit-ins, and letter-writing campaigns and of the political transformation that they have brought about, these media modify climate politics from a spectacle to a site of struggle. Depictions of actual practices of citizen engagement can stimulate empathy and function as inspirational exemplars. By showing how citizens are undertaking concrete political initiatives on climate, alternative media demonstrate that changing politics is feasible and contribute to the normalization of climate activism. In other words, alternative media may be an important agent in the development of agonistic politics on climate change.

In chapter 10, Scandrett, Crowther, and McGregor offer a critical analysis of the exclusion of the voices—and needs—of large parts of society from climate change politics, as well as of alternatives for including them. They show that struggles to define climate change and energy issues have been dominated by specific social identities as the dominant technomanagerial approach to the issue has been shaped by narrow segments of society with specific values, interests, and forms of knowledge. In this chapter, the political in climate politics clearly emerges in connection with issues of class, distribution of power, and social and environmental justice. Scandrett et al. discuss the political consequences of excluding the voices and views of the working classes, the rural and urban poor, and indigenous communities, providing support to Mouffe's defense of agonistic pluralism. The chapter builds the grounds for a radicalization of democratic politics on climate change—communication being critical to the construction of new forms of knowledge and social relations through popular education initiatives of a dialogical nature, as well as to the confrontational politics of direct action, which risk creating new social tensions by challenging dominant structures and powers. The interplay of top-down approaches to public engagement and bottom-up initiatives comes across as often complex and disputed.

The authors of this book examine communication as a key component of social and political dynamics and attempt to demonstrate their interplay through multiple case studies. Citizen groups, from loose "critical communities" to NGOs or even single individuals, such as unaffiliated artists, are vital in the creation of meanings that challenge a given political hegemony. Communicative practices provide the basis for engaging the public, whether through social marketing, public participation, agonistic pluralism, or a combination of all three. The possibilities entailed in that engagement may yet contribute to a transformation of climate change politics that appreciates both the importance of individual political subjects and their communities.

Endnotes

1. As Mouffe (2000, p. 101) noted, "'politics' ... indicates the ensemble of practices, discourses and institutions which seek to establish a certain order and organize human coexistence in conditions that are always potentially conflictual because they are affected by the dimension of 'the political.'"
2. Habermas (1989) has of course also made fundamental contributions to understanding the systematic distortion of the public sphere by the media and how that constrains people's participation in public life.
3. An increase of the global mean temperature by 2 °C until 2100.

Part I

Public Engagement Through Social Marketing

CHAPTER 2

CLIMATE CHANGE POLITICS AND EVERYDAY LIFE

Responses to Climate Communication in Norway

*Marianne Ryghaug
and Robert Næss*

INTRODUCTION: CLIMATE CHANGE AND THE LACK OF PUBLIC RESPONSE

During the past years there has been an increasing focus on climate change and the lack of public response. Although most people are aware that climate change poses a threat and that individual fossil fuel consumption contributes to this threat, few are prepared to change their lives sufficiently to make a difference. As phrased by Anthony Giddens (2009, p. 1): "Why does anyone, anyone at all, for even a single day longer, continue to drive an SUV?" He points to the fact that most people are doing very little, if anything at all, to alter their habits even though they know that these very habits are the source of the threats posed by climate change. In this chapter, we show that Norway is no exception. The country offers an interesting context to study the appropriation of knowledge and sense-making with respect to the climate change problem and climate change politics. With a small population of (about 5 million), Norway is one of the world's largest exporters of oil and gas, on which its economy is quite dependent. Large hydroelectric resources have until recently contributed to a widespread popular perception of energy affluence, whereas the cold climate produces appreciation of access to relatively cheap and abundant energy (Aune, 2007; Aune, Ryghaug, & Godbolt, 2011). This situation, as well as the comfort-oriented energy culture so common in Western societies, could tempt Norwegian citizens to be skeptical of claims about anthropogenic climate change. On the other hand, the climate change threat may raise doubts about the sustainability of the comfort culture.

In this chapter we examine whether the climate threat has contributed to destabilizing the comfort orientation at the household level. Are changes taking place in people's everyday lives, or is there an inconsistency between attitudes and actions? In this way, the chapter offers a diagnosis of the problems that affect people's relation to climate change and a discussion of how they perceive the possibilities of practically addressing the issue. The way the issue is perceived and acted upon is also likely to be influenced by political messages, and the way the problem is dealt with by policy makers in Norway and around the world. Negotiations aimed at limiting climate change by achieving global reductions in greenhouse gases have been organized by the United Nations (UN) since the Rio de Janeiro conference in 1992 and have continued until today with limited success. The problem has reached the political agenda of most world leaders, and most industrial countries currently have some form of climate change politics. However, the way climate change policies are integrated and embedded into national institutions and in the everyday concerns of the citizens will be of utmost importance (Giddens, 2009). In this respect it is important to analyze how the political framings of climate change and the policy tools put forward to mitigate the problem influence the public appropriation processes. This is one of the main focuses of this chapter.

In the first section we provide the theoretical basis for studying the appropriation of knowledge about climate change in an everyday context. We demonstrate that earlier studies have often—and according to us, wrongly—perceived the weak relationship between knowledge and behavior as caused by a knowledge deficit, while failing to stress the importance of the development of practices, the construction of meaning, and processes of learning as important aspects of appropriation of knowledge. The domestication approach that we use in this chapter provides an alternative understanding, taking into account these aspects of the appropriation process. In the following sections we outline the data material and the methods used for collecting and analyzing it, and we describe the national context of domestication, which is important for further analysis. The analysis is organized around two of the sense-making devices that shape the domestication of climate science knowledge: climate politics and everyday life. The last section sums up the findings and discusses the role of media and climate change policy in engaging citizens in the climate change problem on the practical level.

PREVIOUS STUDIES ON CLIMATE CHANGE ATTITUDES, KNOWLEDGE, AND BEHAVIORAL CHANGE

Surveys eliciting attitudes toward climate change consistently indicate the following patterns: high levels of concern, especially for the future; beliefs in specific negative outcomes, such as rising sea levels, more frequent storms, and possible water shortages; and a general agreement that potentially this is a serious problem (Bord, O'Connor, & Fisher, 1998, 2000). This has also been the case in Norway, where 2010 polls portrayed a widespread concern regarding climate change; 70% of the respondents think that climate change is a result of the emission of greenhouse gases caused by human beings.[1] A substantial part of the public tends to relate these changes to their own activities; nonetheless, the same polls demonstrate a preference for a "wait-and-see" attitude concerning actions to be taken. Thus questions emerge about consistency in attitudes concerning climate change.

Several studies have identified numerous misunderstandings regarding climate change and have concluded that these are the reason for people's weak behavioral response (Bord et al., 2000; Bostrom, Morgan, Fischhoff, & Read, 1994; Kempton, Boster, & Hartley, 1995; Sterman & Sweeney, 2007).[2] Internationally, it has been argued that improving the public understanding of climate change science is critical to developing actions and to making acceptable public climate policy (Zia & Todd, 2010). Much of this research has been targeted at detecting flaws in the mental or cultural models of climate change—for instance by looking at problems of cognitive dissonance (Stoll-Kleemann, O'Riordan, & Jaeger, 2001), the trust-gap hypothesis (Priest, Bonafadelli, & Rusanen, 2003), and ideology (Zia & Todd, 2010)—and consequently focuses on the importance of correcting these mental or cultural models of public knowledge of climate change science (Zia & Todd, 2010). In consequence, it has been assumed that there is a need to educate people about this issue and that a better informed public would display more active engagement.

However, the problem is likely to be more complex, and the link between knowledge and behavior may not be as straightforward as assumed. Bord et al. (2000) foresaw three possible trajectories. One possible outcome could be that a deeper understanding of the complexity and the uncertainty characterizing climate change, for example, would make people think that there is little individuals can do besides hope that governments know enough and have the resources and the organizational tools to deal effectively with the issue. Second, more exact knowledge may not be required to stimulate concern or support for environmental actions and policies because concern for possible climate change impacts may be more a function of perceived risk from environmental pollution in general than of a precise understanding of the issue. Third, support for communal problems often follows paths of conformity, so that the low salience that characterizes the climate change issue may reflect a perception that other citizens are not particularly concerned about this issue.

Studies of other environmental problems, like air pollution, have demonstrated that it is difficult to promote behavioral change by enhancing public awareness through public information. Often, the problem has been that the provided scientific representations of the problem have been too abstract and have been perceived as irrelevant by most people (Bickerstaff & Walker, 2003). To overcome this problem, many have called for a bottom-up approach, which could enable people to assess their local conditions and empower them to act responsibly. Without greater sensitivity to the everyday contexts of individual and collective understanding, appeals to individual action are likely to fall on deaf ears (Bickerstaff & Walker, 2003; Owens, 2000). This approach of studying attitudes toward and knowledge about science may be criticized as deploying a deficit model (e.g., Michael, 2002; Norgaard, 2011). Alternative approaches have examined scientists' and lay people's relative expertise, exploring lay local knowledge and the rationality of people's engagement with science (Wynne, 1995). For example, Martin (1994) has shown that the perceived logic of everyday life in a given context is a main feature of the way scientific knowledge is made sense of and accounted for. Members of the public are not passive recipients, excluded from the production and validation of knowledge. They consider, validate, supplement, and adapt the knowledge that is communicated to them (Martin, 1994; Weingart, 1998). Thus in this chapter we explore the appropriation of climate change knowledge in light of the perceived logic of everyday life. One approach based on this perspective that will be fruitful for the analysis of the appropriation of the climate change problem and policy in everyday life is the domestication perspective.

DOMESTICATION AS AN APPROACH TO STUDYING THE APPROPRIATION OF CLIMATE SCIENCE

This chapter analyzes sense-making with respect to climate change by drawing on the generic concept of domestication (Berker, Hartman, Punie, & Ward, 2006; Haddon, 2006). To domesticate means to tame and to cultivate the tamed (Lie & Sørensen, 1996; Silverstone & Hirsch, 1992). The theory was initially developed to analyze the use of new media technologies by households (Silverstone & Hirsch, 1992) and has its roots in cultural studies of media, gender studies of household technology, and consumption studies. It has been expanded to study processes of appropriation of technology, as well as scientific knowledge in many different contexts, including national policy making (Sørensen, 2006).

To analyze domestication of knowledge or information means to study the development of practices, the construction of meaning and processes of learning with respect to a given area of concern like climate change (Sørensen, 2006; Sørensen, Aune, & Hatling, 2000). A main advantage of this perspective is that it reminds one that sense-making is not just about meaning; there are also cognitive and practical aspects that need to be looked at. Phrased differently, knowledge about human-made climate change needs to be enacted in everyday life, and this enactment involves articulation of positions with respect to the truth and falseness of knowledge claims but also considerations about how to act on the perceived challenges. Moreover, domestication may result in rejection or acceptance of climate science knowledge, in addition to a variety of transformations of this knowledge. Rejection or acceptance may even depend on the perceived possibility of sense-making, as well as on the transformation of practices.

The domestication model stresses the reciprocal and simultaneous in negotiations about how to appropriate knowledge. Domestication, then, is a process whereby technological objects or the handling of scientifically described phenomena (like climate change) and the people involved may change. Thus domestication offers insight into changes that take place with respect to human beliefs and actions, as well as with respect to technology and the material environment (Aune, 2007). What is constructed through domestication may, according to Sørensen et al. (2000, p. 241), be understood as "micro-networks of humans, artefacts, knowledge and institutions." This means that what may be domesticated is not just knowledge about human-made climate change but also knowledge about pertinent policy making, political initiatives, and everyday life practices. Presumably, political action also sends messages. In this chapter, we examine how people understand political signals and tools of governance in relation to climate policy, and how are they used as sense-making devices.

Studies of public understanding of science have tended to work from a lay–expert binary in which experts supply the knowledge that the public is assumed to appropriate. As Irwin and Michaels (2003) reminded readers, the "doing" of public understanding is much more complex. The lay–expert and the related science–society binary should be transgressed because the distinction between experts and the public, between science and society is blurred. Moreover, there are many actors involved in the doing of public understanding—media, politicians, industry, public servants, and so on. Consequently, the domestication of climate science knowledge is not a linear relationship between scientists and the public. Media should be assumed to be important in the sense-making of climate change, but in ways that cannot be predicted by a reading of media messages. The consumption or reception of media is a much more complex issue because the actual engagement with media varies considerably (Couldry, Livingstone, & Markham, 2007). At the same time, invoking domestication theory broadens the issue because it highlights the dynamic interaction between knowledge and practice. The political implications of this are related to the need to extend the climate change political agenda. It is no longer possible to distinguish strictly, for example, between policy making and information, or to limit information solely to that about climate science. In line with Hulme's (2009) argument, domestication theory casts climate politics as a politics of everyday life.

Consequently, we anticipate that domestication of the climate change problem and the related issues and knowledge would have quite dramatic effects on the way people think and live. In fact, such domestication would imply quite a radical shift from an already appropriated "hard" fossil fuel–based sociotechnical system with a large number of hybrid associations—such as driving, heating and cooling of houses, and shopping—embedded in compulsory actions to an emerging, softer, and malleable "postcarbon" life. A stepping stone toward this is understanding basic aspects of what causes climate change and the correct responses to the issue. Further, we anticipate that a domestication of the climate change problem means that the problem is recognized symbolically as a serious and pertinent threat to every citizen and that the comfort-oriented energy culture to some extent is met with criticism. Consequently, living "the good life" should no longer be equivalent to high levels of energy consumption. Domesticating the climate problem on a more practical level would mean that one would have to reduce the consumption of energy—in particular, energy from fossil fuels like oil and gas—dramatically, forging a shift to clean and renewable energy sources and using energy-efficient, sustainable artifacts. To what extent may this be observed?

METHOD

We address these research questions with qualitative focus-group interviews. Focus-group interviewing is appropriate when interaction between interviewees is thought to contribute importantly to the information produced (Morgan, 1997; Stewart, Shamdasani, & Rook, 2007). With our focus on sense-making processes, which we assumed would be enacted (at least partly) through discussions about knowledge claims and everyday life activities, we saw focus-group interviewing as a well-suited method; it also allows observations of how points of view may be shared or contested. Of course, this method produces not only data about interaction but also information about the individual participant (Morgan, 1997, pp. 18–22).

The participants in the groups were invited to discuss issues regarding anthropogenic climate change. We conducted 10 focus-group interviews. The participants were recruited through existing social networks, discovered through snowballing. We contacted a person operating in each of these networks, for example, the teacher of a class or the hostess of a maternity group. These people helped with recruitment and the practicalities of the focus-group interview (time, place, etc.). Most of these networks were quite small; thus all members were encouraged to participate and did so unless they were unable to come. When the potential group was larger, as the high school classes were, the interviewees were selected by random drawing. Only one of the social networks we approached declined to take part in the study.

The groups consisted of four to eight people. Most of them knew each other before the interview. This seemed to provide a safe atmosphere, allowing them to speak freely. They were interviewed in familiar places, such as in the home of one participant or in the group's usual meeting place. The setting did not seem to make people strive toward consensus. A few groups were fairly homogeneous in terms of opinions; in most cases there were quite a lot of conflicting views, but such differences did not have any perceptible impact on the level of discussion activity.

In total, we interviewed 62 people: 24 men and 38 women. We achieved diversity in terms of age, for the main age groups were covered, even if there was a predominance of young people. With respect to social background, the participants had a higher level of education than the national average. Still, there was considerable variation along this dimension, as well as in terms of engagement with environmental issues. Of course, this sample does not allow for statistical generalization. For our purposes, in order to identify a diversity of domestication processes it was important that in terms of age, gender, and social background, the most common categories were represented in the data. The overrepresentation of people with higher education could mean a bias in terms of level of knowledge, as well as in the employed strategies of deliberation, but this is of less importance because we do not make general claims about the distribution of such features. Moreover, the group of people with higher education showed considerable diversity with respect to the analyzed issues.

We ourselves facilitated the focus groups; we have extensive experience with qualitative interviewing. Our role was to manage the discussion, to follow up interesting points, and to see that everybody had a say. We also checked that the topics in the interview guide were covered throughout the discussions, but the protocol was flexible enough to allow respondents to discuss other topics that they felt were important. The main items in the interviews were participants' perception of climate change, their main sources of information and how they assessed the information, their views of climate scientists and whether they were in disagreement, the need for action to curb greenhouse gas emissions, and challenges related to energy use. The questions were used to initiate and guide discussions. Usually, the groups debated eagerly and freely in a way that spontaneously led them to cover many of the questions in the interview guide.

The majority of the interviews (eight) were conducted between February 29 and June 6, 2006; the two remaining were done on January 12 and February 19, 2007, respectively. We have not observed that particular events or the increasing focus on climate change issues during this period affected the perceptions of the later groups compared to the early ones in any noticeable way, other than through the examples that were referred to. The interviews lasted about one hour. They were taped and transcribed. We have translated the quotes used in this chapter from Norwegian and tried to retain their oral qualities.

The strategy of analysis was inspired by grounded theory (Strauss & Corbin, 1998). We began by examining the transcribed interviews for salient categories, which were given a label or a code. We then compared these codes to find related subcategories that might be linked to more comprehensive categories. In this process, quotations were selected to represent the various categories and positions as accurately as possible. In the final stage, we endeavored to integrate categories. In doing so, we made use of the generic properties of domestication theory as a basis for making story lines. As categories developed and became richer, we returned to the interview material to search for more examples and to look for perspectives or observations that we might have overlooked in the first rounds.

THE NATIONAL CONTEXT OF DOMESTICATION

The introduction described some traits of the Norwegian energy situation, emphasizing that it is characterized by the dominance of renewable hydropower and a large national income from exported oil and gas. This has resulted in a particular energy culture, affected by the idea that there are abundant sources of energy in Norway. Aune and Berker (2007) demonstrated that "the good life" at home and at work is closely connected to a high level of energy consumption. The use of energy may be seen as a product of a comfort society—a culture in which a comfortable life is expected and taken for granted, in which one is supposed to "allow oneself" some pleasures and to live a practical and convenient everyday life.

Today the comfort-oriented energy culture is challenged by the threat of climate change. Climate science in Norway is well established. The conclusions of the IPCC (Intergovernmental Panel on Climate Change) have been accepted by the government, possibly encouraging public acceptance of the scientific claim of human-made climate change. There is broad consensus among Norwegian policy makers that there is a clear relationship between use of fossil energy sources and climate change, and that the emission of CO_2 is an issue that should be dealt with—the sooner, the better.

Nevertheless, previous studies of the coproduction of knowledge and policy with respect to climate change in the Norwegian context have demonstrated that the impact of climate science on Norwegian policy has taken place mostly on a symbolic level. With respect to political action, the effect of climate change knowledge has been relatively moderate (see Ryghaug, 2011). In fact, policy makers seem to interpret climate change policy as subject to comfort-society standards. Comfort-reducing measures are to be avoided, a view that was clearly expressed by the Norwegian Commission on Low Emissions. This commission prepared scenarios demonstrating how Norway could reduce its emissions of greenhouse gases 50–80% by 2050 and concluded that this could be done with fairly small impacts on welfare and everyday life (Norges Offentlige Utredninger, 2006, p. 18). The commission emphasized technological fixes, including increased energy efficiency and improved technologies for new renewable energy, as the main strategy, stressing sustainable production rather than consumption. "A radical restructuring of the Norwegian lifestyle in a more climate-friendly direction would reduce future emissions much. The committee has not chosen to recommend this partly because we believe it would be an impossible political task to realize" (NOU, 2006, p. 11).

Although there is broad consensus regarding human-made climate change, many politicians have argued that people do not see the gravity of the situation. Norway's former minister of environment Helen Bjørnøy has claimed that the "emerging climate changes we have seen in recent years are simply too pleasant" and that "it is perceived as positive that autumn is getting warmer" (Andreassen, Møller, & Johansen, 2006). This is interpreted as a need to provide more and better information about climate issues. Accordingly, in 2007, the government launched a national climate campaign. The need to educate the people has also been articulated by other Norwegian politicians, such as the former leader of the Liberal Party, Lars Sponheim ("To save the environment, each one of us has to do something in our everyday life") and Magnhild Meltveit Kleppa, presently minister of transport, who has said that "environmental problems are so important that we all have to do something about it" (Johansen, 2006). However, the current minister of international development and environment, Erik Solheim, has claimed that environmentally friendliness must emerge as a simple and profitable set of actions.

Consequently, the context of the sense-making with respect to climate change among members of the Norwegian public should be considered complex and contradictory. To begin with, there is broad agreement that climate problems are human made and that immediate action is needed. Some policy makers claim that the problems will be solved by technical fixes and that the everyday life of most people will not need to change much. Accordingly, measures like increased investment in research and development regarding new renewable energy and carbon capture and storage have been put in place. Others argue, as already mentioned, that the general public has to take greater responsibility. In sum, most politicians seem to wait for changes in public opinion as a basis on which to enact a more concerted and active climate change policy.

Making sense of climate change

When analyzing the focus-group interviews, we found that the domestication processes could be fruitfully understood by reference to what we call sense-making devices. A sense-making device is a particular form of framing for an issue. We found that the domestication of climate science knowledge among the Norwegian public was shaped through the use of mainly five sense-making devices (Ryghaug, Sørensen, & Næss, 2011): (1) media coverage of dramatic events in nature, including extreme weather, (2) news media's enactment of a controversy between climate scientists and climate skeptics, (3) the practice of critical readings of media's messages, (4) observations of political action, and (5) reflections with respect to one's own everyday life. In accordance with our research questions, we here focus on the latter two sense-making devices: climate politics and everyday life. However, before moving on to these two, we quickly review the first three. They all relate to the fact that news media accounts of climate change issues are important to people's perceptions, but not in a straightforward way.

In general, the focus-group participants stated that news media, above all newspapers and national television, were their main source of information about climate change issues. This finding aligns with Norwegian media surveys. According to Ryghaug (2006), in the period between 2002 and 2005, major Norwegian newspapers drew upon two broad narrative strategies in their coverage of climate change issues. One characterized the state of knowledge about anthropogenic climate change as a scientific controversy. This finding is similar to studies of "balanced reporting" (Boykoff & Boykoff, 2007). The other strategy was to use dramatic natural phenomena like extreme weather to provide a kind of public proof of climate change (see also Olausson, 2009; Weingart, Engels, & Pansegrau, 2000).

Focus-group participants made reference to both strategies, but in diverse ways. Some read the enacted controversy as a reason to be skeptical, some used it as an argument to moderate their perception of how serious the problem was, and others saw the controversy as media made. News media's invitation to use dramatic events in nature as a public proof of climate change was accepted by many, whereas some interpreted the events differently, using their own experiences to claim that there was no climate change. Still, this sense-making device seemed helpful to most of the focus-group participants as a tool for domesticating climate change knowledge. This device made climate science knowledge appear relevant and assessable because it linked scientific arguments about human-made climate change to observable and understandable phenomena like extreme weather or weather changes.

In contrast, the media narratives with respect to nature were often modified in the reception process and given a more alleviated, less agonizing meaning. The ensuing sense-making of climate change by many interviewees resulted in efforts to diminish the risks involved: The situation was grave but not as severe as the media would have it. One line of reasoning was to argue that the climate problem was distant in time and space, thus developing a locally framed and less stressful understanding. For example, some of the interviewees saw the real frightening drama happening (or going to happen) in far-away places or in a distant future. Quite a few thought that climate change would not have critical consequences for people living in Norway. Moreover, in general, the focus-group participants were quite reserved about the accuracy of media accounts and argued the need for critical interpretation.

The perception of climate change problems as distant from everyday concerns and of secondary importance, making climate change less imminent than other problems, is similar to what Norgaard (2011) found in her ethnographic study of a small community in Norway and to what Lorenzoni and Pidgeon (2006) observed in the United States and in European countries. "Public concern for climate change appears to be tempered by uncertainty about whether and when climate change will occur, the degree of change and by competition from other seemingly more relevant issues of individual concern" (Lowe et al., 2006, p. 438). This point has also been stressed by Giddens (2009).

Most of the focus-group participants had a fair understanding that there was a human-made climate change problem related to emissions of greenhouse gases, or at least knew that this was a widespread claim. Generally, participants had domesticated the knowledge about climate change, above all cognitively, by accepting that climate change was happening at least partly owing to human action. This is widespread, as indicated by a recent representative survey of Norwegian citizens in which 69% of the respondents agreed that the climate change problem is a very serious or serious problem (Karlstrøm, 2010). There was little demand for more knowledge from climate scientists about the dynamics of climate change. What was asked for was more information regarding what they themselves could do about the problem. Such information was perceived as less accessible. As one respondent explained, "I think we have had enough information about the fact that the there is a crisis and that the earth is in trouble, but there is not enough information about what *you* can do." How could climate change knowledge be domesticated in order to provide a basis for individual action?

Climate change appropriation and political enactments of the problem

We now turn to observations of politics as a sense-making device. It emerged from the interviewees' accounts of political action—or rather, their accounts of the lack of such initiatives. Many focus-group participants were critical of what they saw as passive politicians, whereas others were puzzled by what they saw as a contradiction between the scientific claim about a very serious global problem and the lack of visible political action to solve the problem. Because, in principle, politicians could manage truly serious challenges, doubt emerged when politicians were seen not to engage in radical, large-scale initiatives to, for example, curb CO_2 emissions. Lack of visible action was seen to support the belief that the impact of climate change was exaggerated by climate scientists as well as by the media. Liv stated, "I do not think it [climate change] is human-made only—I don't—because if it had been, I think that the governments in various countries would have done something!" This confirms the previous assumption that the domestication of knowledge of climate change may be influenced by the way the problem is acted upon (or not) by politicians. Lack of political action was also interpreted by some as a question mark with respect to the reliability of scientific knowledge. Arguably, these participants observed a failing coproduction of knowledge and politics (Jasanoff, 2005), a lack of reinforcement between science and policy to achieve stability of both these sides of the equation.

Some participants also pointed to what they saw as inconsistencies in present social developments—for instance, regarding infrastructure and modes of transportation that affected their everyday lives. For example, some perceived the increasing availability of cheap airline tickets or the support of motor sports as inconsistent with the serious messages from climate scientists. John argued:

> I think that most people try to do their best. However, society is often organized in ways that make it counterintuitive to act environmentally friendly. It is not arranged for, ... with the cheap plane tickets and ... I can take the bicycle to work and to the kindergarten if I want to, you know. But consider these banal things such as the emergence of cut-rate airlines, for example, and that they shut down the night train between Oslo and Stockholm, and a bunch of such obvious things—where there are no alternatives to CO_2 emissions.

Thus people felt that they received mixed signals when, for instance, airplane tickets were extremely cheap but more environmentally friendly alternatives like trains were cut out. There was a widespread impression that the authorities did not do enough in order to counter climate change and thus could be seen to support the view that climate problems were not that pressing. Underlying this was the—perhaps romantic—expectation that problems of such magnitude as climate change would be dealt with by industry and government. Wahid put it bluntly: "It is big industry and governments that may do something. We [ordinary people] cannot really do anything. To turn off the light doesn't help." Some thought that Norway's general level of prosperity and dependence on oil and gas represented one of the biggest problems related to climate change. On this basis one questioned the political will to actually do something about the climate problem, and some interviewees noted the paradoxical situation in which the authorities aimed for continued large-scale production of oil and gas (even opening up new oil fields) at the same time as they requested the citizens to save energy, drive less, and buy climate quotas when flying.

Thus politicians and public authorities were blamed for the lack of problem-solving action. The interviewees asked for consistent governance. They wanted politicians to be more sensitive to the claims of the scientists concerning climate change. At the same time, people disliked the moral finger-pointing of the politicians that also was perceived as unhelpful in the organization of everyday life. We observed several discussions in the groups about whether they should feel guilty about their actions (or lack of actions). If someone else (particularly politicians and public authorities) were to blame, why should ordinary people have to feel guilty for the problems? Observations of a lack of political will to do something about the problem (both nationally and internationally) also fostered a sense of powerlessness among many of the interviewees. This sentiment was expressed by a young man in the following way:

> It is quite depressing to hear reports from the top political level. It is a success that they have agreed to meet again in three years to talk more. It is on that level. It is a parody, at least the reports that one gets from these big conferences, but perhaps there is a process going on secretly, that I do not see.

Many interviewees observed a lack of comprehensive and binding agreements on reducing greenhouse gas emissions from the international community as a setback. Arguably, there seems to be an impasse between the public and the politicians. The interviewees asked for political leadership, to be convinced that climate change is a pressing and serious problem. What they observed was politicians waiting for public opinion to strongly favor actions to curb emissions of CO_2 before taking action. This situation was clearly not facilitating the domestication of the climate change issue into everyday life.

THE APPROPRIATION OF THE CLIMATE CHANGE PROBLEM IN PRACTICE: WHAT TO DO?

The second sense-making device was observed while analyzing interviewees' reflections about everyday life issues, in particular on the problem of how to act regarding challenges related to climate change. Most of them said they felt fairly powerless in the face of what they saw as insurmountable challenges. Besides engaging in recycling as they were asked to by the government and perhaps also saving a little on energy use and car driving, most focus-group participants did not see how their individual behavior could matter. Some used this as an excuse to continue with their present way of life. Others, like Eskild, stated more bluntly, "I don't think any of us around this table is willing to reduce their standard of living." Thus the domestication of climate change knowledge may be difficult because it appears to be dissonant with the established practices of everyday life. These kinds of difficulties of integrating climate change into everyday life are also described in Norgaard's (2011) study of a Norwegian community. In the survey previously referred to (Karlstrøm, 2010), many respondents (60%) claimed to have changed their way of living as a result of the climate problem. When asked what they did, the most common response was that they reduce their energy consumption (50%), recycled their waste (39%), and used public transport more frequently (29%).

Domestication theory allows a greater emphasis on practical and material dimensions related to infrastructure and technologies. From the interviews it was evident that access to alternatives to established patterns of energy use and modes of transportation was a key aspect of the domestication process and often made it difficult to translate climate knowledge into practice. Many expressed a sense of powerlessness, which may be interpreted as a frustration stemming from a perceived lack of capacity to act—a sense of limited agency:

> Kathrin: It is very difficult to know how one should live in order to live and to act environmentally friendly. If driving is so wrong, why is it permitted? Why do we have petrol cars?
> Ann: In that respect the authorities have a job to do.
> Kathrin: I agree. We must be allowed to know what we are supposed to do.
> Ann: And it is difficult to sort out—what is important and what is not....
> Kathrin: It is not sufficient that they tell us, "Limit your car use and save energy!"

Note Kathrin's question, "If it is so wrong to use the car, why do we have them?" This demonstrates that current policy and lack of regulation may easily be interpreted to mean that car use is not a serious problem. Moreover, many interviewees thought that they were doing what they could in order to counteract climate change. Nonetheless, it was clear that they saw it as difficult to live a low-emission life within a comfort society that offered little room for proenvironmental actions.

The overall impression was that focus-group participants saw Norway's energy situation as unproblematic. Only a few related their own energy consumption to the national energy situation and to climate change. When confronted with the issue, many were eager to stress that their own consumption was low compared to the consumption of others, implying that it did not matter what they themselves did.

Increasing standards of living have made it common to use energy for many purposes seen as indispensable, such as electricity for lights, laundry, and the heating of homes. Such energy consumption was taken for granted. When our interviewees considered energy use as a problem in light of climate change, they thought of industry or luxurious consumption:

> Returning to what I said earlier about having a bad conscience for letting the car run idle: When watching Rally Norway, I thought, "What the heck—does it matter if I let my car run idle for three

minutes when they are using four liters of petrol every 10 kilometers?" There you have this gang of people that are playing around out in the woods.

(Kristin)

The activity that most commonly came out of the domestication of the climate problem was recycling of waste. However, even this activity brought out some doubt and added to the feeling of powerlessness, as it was considered frail. According to Tanja, what she could contribute was

> little things, like things that do not take too much time. Everyday life is so busy, and large efforts are needed; I'm not so sure that many will bother. If small matters may change something, then one can take part in it.

This quote suggests that climate change challenges may be interpreted within a comfort society as a frame of thought. One may be willing to act according to environmentally sound principles as long as it is easy to do and does not take much time and effort. However, the main impression was of a weak link between one's own energy use and energy production as a cause of climate gas emissions. The reflections were mainly moral:

> Honestly, I must admit that I have bad conscience when I start the car and let it run idle before I scrape [the ice] off the windscreen. I have. And normally I scrape the window first and start afterwards. I am doing that with a view to climate change, you know, or the ozone layer. Apart from that I do not know if I am doing so much wrong.

(Kristin)

One is struck by the concern for running the car idle for three minutes, whereas driving a car is not addressed as a challenge. This was typical. It was particularly when consumption was extreme or regarded as highly superfluous that it was seen as wrong. It was also common to make comparisons between the privileged and the less privileged, between those with wasteful and extravagant behavior and those whose behavior was more sober-minded. In fact, there was a great deal of deliberation concerning issues of justice: who should pay for mitigation activities, who should be allowed to drive cars, what kind of vehicles should be allowed, and how much driving should be permitted. Underlying this were concerns about responsibility, as well as its distribution. Should rich people be allowed to pay their way out? Consider the following exchange:

> Eirin: It is feeble I feel—it is like "remember to turn off the light and lower the temperature during the night."
> Linn: Then I actually think more about the electricity bill, than...
> Eirin: Yes, I am absolutely motivated by it.
> Katrin: But at the same time there are people with single-family houses of 200–250 square meters—they use far more than those of us who have small apartments. And if the light is on in the bathroom through the entire night, it doesn't matter.

According to Karlstrøm (2010), almost 80% of the Norwegian population thinks most other people are wasting electricity.

Many interviewees expressed the idea that a Norwegian citizen has the right to use a certain amount of energy and that this amount may be used the way one wants, as if everyone had his or her own invisible electricity quota. The size of this quota would vary according to life situation and life stage. For example, energy consumption related to ordinary, necessary daily routines, such as driving the kids to kindergarten or the heating of homes, was considered unproblematic. However, the widespread sentiment that ideally everyone should have the same quota of energy could be seen as aligning with the typical Norwegian emphasis on social equality. In this way, it became morally defensible to drive an extremely fuel-inefficient car during the summer if you used public transport the rest of the year.

Thus the domestication of climate change knowledge with respect to the actions of everyday life is a complex achievement. First, the climate change problem must be acknowledged as important in the context of one's own everyday life. It has to be acknowledged as an issue that one must actively deal with. Second, it has to be domesticated into practice by people adjusting their own behavior so that they take up actions that they feel will contribute to mitigating the problem. Third, in order for the climate change problem to be domesticated, one has to address issues related to a fair distribution of benefits and loads. Social justice appears to be important to the acceptance of regulations and measures meant to mitigate climate change. Our analysis shows that if climate change policies are to be integrated and embedded in everyday life, they must be perceived as fair and the burden must be seen as collectively distributed. Thus, all things considered, we see that reflections on everyday life represent a fairly comprehensive way of making sense of climate change.

CONCLUSION

In this chapter we have described how Norwegians domesticate the climate change problem and policy and how this affects their actions —for instance, their use of energy—in everyday life. In particular, we have focused on sense-making devices and the construction of perceptions, attitudes, and practices at the crossroads of a dominant comfort culture and an increasingly strong message of anthropogenic climate change. Clearly, the climate change issue does not suffer from a knowledge deficit, although more information about how to mitigate climate change has been requested. The interviews indicate that on the whole, people had fairly good knowledge about the main features of human-made climate change. This does not mean that everybody accepted the claims of climate scientists, politicians, and news media. As anticipated, the interviews show that climate knowledge was critically evaluated in ways shaped by the interviewees' assessment of climate change politics and their own everyday life situations, in particular, their perception of their agency or ability to act on the problem.

With respect to symbolic aspects, the domestication of climate change was complex. We observed a widespread belief that climate change should be a cause of concern and worry. At the same time, climate change issues were fairly absent from everyday life considerations. Largely, they were perceived as distant in time and space, as shown in many earlier studies (Giddens, 2009; Lorenzoni, Nicholson-Cole, & Whitmarsh, 2007), and less pressing than many other challenges. As in other studies about how scientific knowledge is integrated in daily life (Dierkes & von Grote, 2000; Irwin & Michaels, 2003; Norgaard, 2011), the main effort of our interviewees was to make climate change knowledge relevant to their own lives. However, domesticating climate change knowledge into practice was more difficult. The focus-group participants found that besides doing "the small things," like engaging in recycling and using their cars a little less, it was difficult to figure out what they ought to do. The resulting widespread feeling of powerlessness in relation to the climate change problem meant that it was hard to domesticate and embed in everyday life by generating relevant practices. In many cases, doubts prevailed about whether there really was anything that they could do to help mitigate the problem. This meant that the domestication of climate change knowledge was hampered by a lack of perceived relevance. It was difficult to transform climate change awareness and knowledge into actions other than worrying.

Political actions were not a positive sense-making mechanism, either. We observed that many interviewees considered appropriate political action to be proof of the seriousness of the climate change issue. As noted, there was a failing coproduction of knowledge and politics to achieve a stable acceptance that anthropogenic climate change is a very serious problem. In addition, there was a widespread frustration about inconsistent messages with respect to how the problem would be solved and by whom. Still, many voiced a moral anxiety: After all, maybe they were not doing enough of the right things.

Previous research often holds news media responsible for the lack of public belief in human-made climate change and the lack of support for mitigating action. News media are accused of exaggerating climate change controversy and of engaging too much in scare tactics (Hulme, 2007; Revkin, 2007; Ryghaug & Skjølsvold, 2010). We found that news media played an important but contradictory role. In their coverage of climate change, they produced two sense-making devices that were employed in the domestication of climate change knowledge (Ryghaug, 2006). One was the use of unexpected events in nature, like extreme weather, as proof of human-made climate change; the other was the presentation of climate change knowledge as controversial. Owing to the widespread skepticism with respect to the reliability of news media, the third sense-making device, the media's narrative strategies were read in complex and contradictory ways. Though there is no doubt that news media are the most important sources of knowledge of climate change, there are nevertheless other factors to take into consideration.

We have found two such factors or sense-making devices, as outlined. First, it seems that scientific knowledge may be validated in an important way by being put to political use. Our interviewees asked for political leadership in order to be convinced. In this respect, they were disappointed. We found that the domestication of climate change knowledge based on perceived political inaction produced a tempered view, namely, that human-made climate change might be a fact but is not as pressing as scientists claim. Second, climate change knowledge is assessed on the basis of everyday life experience. Knowledge that challenges the way one lives seems to encounter resistance. As we have seen, it is tempting to dismiss or moderate such knowledge. This may be reinforced by difficulties encountered when trying to transform climate change knowledge into doable action. Many interviewees complained about what they saw as a lack of opportunity to do the right thing.

The latter problem may—at least to some extent—be amended by providing more and better information about relevant individual activities. However, because the observation of relevant political interventions seems to be so key to the validation of the related knowledge, in our case about climate change, the lack of such interventions becomes critical. Thus what we see is a clear relationship between climate change policy or political action or inaction and everyday life. This is particularly obvious in relation to infrastructures and technologies in which the availability of practical alternatives to uses of energy or modes of transportation largely depends on political will and affects everyday life. For example, people are asking for mitigation policies that will change the infrastructures that shape the way they live their lives. This includes comprehensive policies for electric road transport, better and cheaper modes of public transport, political guidance concerning energy efficiency in buildings, and willingness to develop renewable energy technologies. It seems quite clear from our interviews that neither news media efforts nor governmental information campaigns can overcome the negative effects of a lack of visible political action. When policy making, in addition, provides inconsistent arguments, it naturally makes matters worse. The contradiction between arguing that the climate problem may be managed through near unnoticeable technical fixes and asking that the public take on the prime responsibility to reduce CO_2 emissions is at best confusing and seemingly deterrent of deeper citizen engagement and behavioral change.

Endnotes

1. This is 4% less than the previous year, according to TNS Gallup's latest climate survey from 2010 (TNS Gallup, 2010).
2. For instance, Americans have tended to underestimate the role of automobiles, utilities, and home heating and cooling in emitting large quantities of CO_2, and many people have erroneously believed that aerosols, insecticides, nuclear power generation, and depletion of ozone in the upper atmosphere exert significant impact on climate change (Bord et al., 2000).

CHAPTER 3

LANDSCAPE-BASED DISCOURSE AND PUBLIC ENGAGEMENT

COMMUNICATING CLIMATE CHANGE IN U.S. NATIONAL PARKS

*Sarah Schweizer and
Jessica Thompson*

INTRODUCTION

Nearly 30% of the United States is designated federally owned and managed public lands, and with this privilege comes a political opportunity to communicate the impacts of climate change to the public. The U.S. National Park Service (NPS) has the power to effectively disseminate information and engage audiences through in-depth landscape-based experiences and discourse. In 2010 the national parks (hereafter parks) captivated the attention of 278,939,216 visitors enthralled with majestic landscapes and historical-cultural treasures (National Park Service, 2012a). The parks have a capability to facilitate citizen engagement and inspire political action through the specific climate change messages they provide visitors and the mitigation actions they recommend. Visitors can be engaged as citizens responsible both for influencing climate change policy through elections and for making lifestyle changes directly related to reducing greenhouse gas emissions. Parks also engage visitors as they discuss the role of the parks and the federal government in tackling climate change. However, many federal managers and park rangers avoid discussing climate change. Climate change messages developed by the parks will be important in determining the saliency of the issue for visitors and will influence their perceived response efficacy and self-efficacy. The parks offer an ideal venue to inform visitors, influence perceptions, empower behavioral change, and facilitate public pressure on political leaders to support climate change policy; but park authorities appear to be missing this powerful opportunity because of the political implications they face as a federal agency.

In this chapter we explore ways three of the most iconic U.S. national parks are communicating about global climate change through a generative analysis of a combination of artifacts. Through our analysis we argue that parks offer a largely unrealized opportunity for landscape-based discourse about climate change. Landscape-based discourse is our proposed conceptual framework for making sense of public communication about climate change in specific and unique landscapes. Based on what we know about place-attachment theory and insight from literature on place-based climate change communication (largely influenced by theories of experiential learning), we build a foundation for understanding the rhetorical power of parks to communicate climate change. This chapter outlines a critical framework for understanding how parks in the western United States currently communicate about climate change—specifically, whether they utilize a landscape-based approach. Ultimately, we seek to understand how a physical, federally managed space can function as a forum for public engagement of climate change through imparting a deeper understanding of the ecological impacts and possible management responses to climate change.

THE CHALLENGE: COMMUNICATING CLIMATE CHANGE AND ENGAGING CITIZENS

Numerous writers have described climate change as one of humanity's greatest challenges (Silver, 1990; Speth, 2004). Many factors have explicitly challenged the effective communication of climate change science to the public. First, there is an enormous time lag in the change in climate and changes in our social system; this is coupled with the assumption that the impacts of climate change most directly affect the developing world (Moser & Dilling, 2004). Second, there is a widening gap between the public's awareness of what action is needed and what actions are being taken. Without an understanding of what to do, individuals are left feeling overwhelmed and frightened, or they blissfully ignore the magnitude of the issue through denial (Moser & Dilling, 2004). In addition, Maibach, Roser-Renouf, and Leiserowitz (2009) identified six target audiences within climate change issues in the United States. The distinct audiences range from alarmed to dismissive, varying in level of concern and action toward climate change. This diversity forces climate change communicators to tailor messages to meet the needs of each group's beliefs, perceptions, and understanding of climate change impacts.

Complicating this lack of understanding is the problem that when climate change is reported in the news, it is often accompanied by images of weather disasters. From earlier research (i.e., American Psychological Association Task Force, 2009; Trumbo, 1995) it is known that the public understands weather and natural disasters as "acts of God" and often fails to see that individual actions influence the pace of climate change. Overcoming this challenge requires that climate change communicators connect human choices and behaviors to the cause of climate change events by educating their audiences on the complexity of system dynamics.

Another potential challenge in communicating about climate change is that the message has no single or uniform voice. There are hundreds of well-organized groups and agencies with agendas for dealing with climate change, but it seems that this polyvocal public sphere has facilitated more polarization than understanding. To overcome this challenge, organizations and agencies should seek opportunities to collaborate by creating clear, concise, and consistent messages about climate change.

The ultimate challenge in communicating about climate change with the public is that few local examples or illustrations argue that it is happening now and affecting current life and landscape. The perception of climate change is that it is uncertain, controversial, far in the future, and out of the public's hands (Leiserowitz, 2007). We argue that these perceptions could be modified through a discourse that emphasizes context, provides an ecosystems-based explanation, and identifies specific actions that the public can take—today—to slow the impact of climate change. This approach points out actions that are accessible and easy to maintain and that empower individuals with a sense of personal efficacy (O'Neill & Nicholson-Cole, 2009). Messages with context and an ecosystems-based explanation would highlight the changes and impacts observed at a specific site, identifying and sharing insights of the changing parks and thus facilitating an inspiring public discourse about climate change at politically designated sites across the country.

Scope and approach

We collected and examined climate change artifacts and communication techniques used by the parks through a critical generative analysis. Generative analysis allows the flexibility to generate units of analysis based on relevant theory. This analytical process focuses our lens and illuminates significant features of the artifacts, allowing us to critically answer our research question: How do national parks in the United States currently communicate about climate change—are they utilizing a landscape-based approach and maximizing this political communication opportunity?

We chose to focus on three national parks that are revered as wonders of the western United States: Yellowstone National Park and Rocky Mountain National Park are both ranked among the top-ten visited National Parks (National Park Service, 2008a), and despite its lower attendance, Glacier National Park was essential to our study because of the unique influence of climate change on resources within its boundaries. There are few places where communicating climate change is more relevant than in a park that is home to more than 150 glaciers. There, unlike in most parks, the melting glaciers confront visitors, forcing them to viscerally experience and comprehend the impacts of climate change. There is no denying what is happening right in front of their eyes. The landscape is an opportunity to connect the visitors' behavior to the impacts they are experiencing in the park.

This study is an analysis of artifacts collected and observed at these three parks during extended visits between May 2008 and August 2008 and again in June 2009. We began by experiencing and interpreting many of the messages a typical visitor would be exposed to while touring these parks: ranger-led programs, photographs, films, tours, brochures, books, and exhibits. We also propose an analysis of the landscapes themselves—the glaciers, animals, water cycles, plant species, and other ecological signs—as material evidence that has the potential to influence the tourist's interpretation of climate change within that landscape.

The primary purpose of this analysis is to identify how messages and landscapes of the parks may influence visitors' understanding and perception of the importance of climate change, as well as confidence in their ability to take action on the issue. Federally protected landscapes like the parks represent a unique opportunity to empower visitors and citizens with a sense of personal engagement with climate change policy and actions through a visceral, material, and emotional experience. We begin our project by asking whether national parks are utilizing this unique communication and engagement opportunity.

Landscape-based discourse

Landscape-based discourse about climate change provides an opportunity to link the impacts of climate change on specific landscapes to human choices and behaviors. Through this connection, communicators have the opportunity to inspire behavior change by considering the audience's connection to place and ability to engage in civic discourse in a place-based context. We propose this conceptual framework to make sense of the artifacts we have collected. This framework allows us to analyze multiple dimensions of climate change communication and argue for the potential power of landscapes to tell the story of climate change. With carefully crafted messages, park representatives (or stewards of any landscape) can: (1) illustrate the impacts of climate change by emphasizing impacts in the immediate local context, (2) connect climate impacts to human behavioral choices through systems-based explanations, and (3) provide concrete suggestions for specific actions, thus overcoming the typical challenges of communicating about climate change.

Connection to place

What happens to one's connection to place when climate change impacts are seen and valued landscapes begin to change? What happens when people no longer feel the same connections and meanings to treasured or inspiring landscapes? These are legitimate concerns that could influence a person's understanding of climate change. It would be beneficial for climate change communicators and park managers to understand the bonds and different forms of attachment that people have for landscapes (Williams & Vaske, 2003). Extensive research has covered place attachment in public space and in nature and wilderness experiences (Bricker &

Kerstetter, 2000; Low, 2000; Steel, 2000; Vitterso, Vorkinn, & Vistad, 2001; Wickham, 2001), but place attachment alone cannot explain the meanings people place on landscapes and how those meanings are altered as environmental crises arise. Stedman (2003) pointed out the chance to take advantage of peoples' bonds to specific places: "Place-protective behaviors are especially likely to result when attachment and satisfaction are based on preferred meanings that are threatened by potential changes to the setting" (p. 567). Moore and Scott (2002) also discovered in their Cleveland, Ohio, study that residents are more likely to become active opponents if they sense that what they value about the place or landscape is at risk. Furthermore, O'Neill and Nicholson-Cole's (2009) study on visual and iconic representations of climate change reinforces the importance of local connections in the communication of climate change:

> All groups made it clear that local impact images are necessary to make people feel empowered to make a difference. They also insisted that a global context should be included, to make the seriousness of the issue resonant, though this should be done carefully so as to avoid making people feel afraid or overwhelmed and totally helpless.
> (O'Neill & Nicholson-Cole, 2009, p. 374)

We propose that connecting the appreciation for a specific landscape with an individual's ability to learn in a place-based context may empower visitors while educating them about climate change in the national parks.

PLACE-BASED CLIMATE CHANGE COMMUNICATION

The practice of learning outside has been called many names, including bioregional education, environmental education, outdoor education, place-based education, and experiential learning. Despite different labels, these concepts are often interconnected and have similar meanings. For the scope of this project we focus on place-based climate change communication as a type of experiential learning and a type of public engagement at national parks because of the opportunity to link climate change impacts with the landscapes that millions of visitors treasure. Place-based climate change communication depends on connecting people to the land through applied learning and firsthand experiences in the field. Relevance is the key to engaging visitors in a learning experience within one of their favorite landscapes. Thomashow (2002) pointed out that the most effective way for people to understand and learn about the changes in the environment is by developing an intimacy with the land around them.

Much of the current place-based communication research focuses on children's learning experiences, but we believe that the underlying principles are applicable to audiences of any age. It is important that park visitors be encouraged to understand and appreciate natural environmental processes before they try to digest the complexity of global climate change and make appropriate behavior changes. Sobel (2004) observed that "authentic environmental commitment emerges out of firsthand experiences with real places on a small manageable scale" (p. 34). "What's important is that [people] have an opportunity to bond with the natural world to learn to love it, before being asked to heal its wounds" (p. 9).

Sobel's research reinforces the importance of developing landscape-based discourse opportunities and the political implications of fostering the public's love for the natural world. For example, parks offer a unique opportunity to facilitate citizen engagement with climate policy by fostering an individuals' connection with nature so that they will feel empowered to protect that landscape.

Analyzing Climate Change Communication at U.S. National Parks

We collected and analyzed a variety of artifacts in order to discover how national parks are communicating about climate change and whether they are using the full potential of a landscape-based approach. Artifacts included brochures, books, videos, interpretive signage, interpretive program descriptions, newsletters, and other text- and Web-based documents about the park and the effects of climate change in that region, or more appropriately, in that ecosystem. Critical examination of the artifacts allowed us to identify relationships and patterns linking the whole set of artifacts for a park, as well as the potential for the park to affect public discourse about climate change. We engaged in an iterative and generative critical analysis, in which we continually identified key terms and themes, revisited our codes, and generated specific units of analysis appropriate to the context and meaning of the message situation at the parks. The foundation for our critical framework included such concepts as landscape-based discourse, connection to place, and place-based education, which all informed our units of analysis for critically examining the collection of artifacts.

Through open coding we identified and analyzed common climate change themes across various forms of messages (i.e., brochures, ranger talks, interpretive boards) in the parks. We began with an open reading and open coding of all of the materials collected at the site (e.g., brochures, videos, program descriptions, newsletters); then we compared open codes, discussed, and negotiated to reach five main coding categories: (1) scientific information about climate change, (2) climate change impacts on the local landscape, (3) social and management aspects of climate change in the park, (4) opportunities to include climate change in place-based education messages, and (5) sustainability initiatives and green management efforts. We agreed that these codes captured the main messages about climate change presented in at the parks. Table 1 lists and describes the coding categories and provides an example of each.

Our goal was to analyze each park as a complete artifact; we read each park's communication as well as its landscape in order to determine whether a landscape-based discourse was being used and whether such an approach would be useful in engaging visitors in public discourse about climate change at that site.

In determining whether a landscape-based framework was used, we analyzed the park's messages for evidence of appeals to one's connection to the place and messages designed with an understanding of place-based education. Through this process we identified examples of landscape-based discourse. These examples are characterized by three main principles: (1) emphasizing the impact of change on the immediate landscape, (2) providing a systems-based explanation connecting human behavior and landscape change, and (3) providing specific actions that the public can take to mitigate climate change while connecting to a desire to preserve the park's landscape. The open codes, combined with the three principles of a landscape-based discourse and informed by theory, provide a framework for interpreting the political opportunity to change the conversation about climate change at three of America's favorite national parks.

Rocky Mountain National Park
Rocky Mountain National Park (Rocky) has been one of the most visited parks since its creation in 1915. Over 150 million tourists have been drawn to the majestic Rockies, now spanning over 265,000 acres (NPS, 2008a). Ecosystems range from peaceful montane meadows to the harsh elevations of alpine tundra.

Table 1. Open coding categories of communicating climate change at three national parks.

Category Label	Initial Coding Notes	Description	Examples
Climate Change (general)	General info on climate change Scientific facts Global statements about climate change	Writing about warming trends and precipitation – not necessarily in the context of the park, but possibly the nation/region	"The earth has experienced fluctuations in temperature and climate with extremes of glacial ice and extended periods of warming & drought. Human activity is now playing a role in these fluctuations." (National Park Service, 2008b)
Climate Change Impacts on Local Landscape	Information about climate change impacts on flora, fauna, ecosystems, natural systems, glaciers	Detailed and scientific information specific to the area or region of the National Park Based on science that has already been conducted/completed about climate change in the area.	"The Park's glaciers are shrinking, says Visty, but not at the alarming rate of those in Glacier National Park to the north. Rocky Mountain National Park's signature alpine tundra is at risk as the underlying ice, called permafrost, melts." (National Park Service, 2008f)
Communication and Management Aspects of Climate Change in the Area/Park	Management of science Management decisions Managing for climate change Mitigation options Education Communication Policy implications	Specific information about that Park's management of science and climate change data collection, as well as management decisions and public communication and education about climate change in the Park or regional area	"Park managers will also have to consider how to interpret climate change for their visitors." (National Park Service, 2008b) "Budgetary constraints will likely prevent the park from single-handedly undertaking more than a few of the research and monitoring projects important to understanding climate change impacts." (National Park Service, 2008b) Action words: "identify, study, monitor"
Opportunities to Include Climate Change in Place-based Education Messages	Landscape-based educational messages Pine beetles infestations in lodgepole pine forests	Messages about the landscape – without climate change included Opportunities to include climate change science relative to that landscape	"Ecosystem of the Rockies" – detailed information on front page of Rocky Mountain NP brochure distributed at Park entrances. "Moraine Park Visitor Center... Interactive exhibits on the past and present landscape, and a bookstore." (National Park Service, 2008g)
Sustainability / Green Management Efforts	Green initiatives Sustainability Green management Eco-friendly Climate Friendly Parks	Messages about efforts to operate the park in a more sustainable manner – could be linked to climate change mitigation messages	"The Park is actively engaged in green practices, including using many hybrid/alternative fueled vehicles, a bicycle for mail delivery, low-wattage compact fluorescent light bulbs and active recycling of office materials." (National Park Service, 2008g)

We were greeted with a heavy mountain rain when we arrived at the east entrance visitor center. Our first priority was to examine every interpretive sign and scout the scene for messages about climate change. We started by asking the ranger there whether he knew of any places where we could see evidence of climate change impacts in the park. He replied that no information was available. He continued by explaining that there was a great deal of speculation as to what causes climate change and regarding the direct effects of climate change. The ranger said he had heard that people blamed climate change for the elk's calving earlier and the pine beetle infestations. He stated that there was no evidence that glaciers were retreating in Rocky Mountain National Park. Appearing uncomfortable with the subject, he quickly left us to answer other visitors' questions. We listened as they asked which hiking trails would not be muddy and which trails they could complete in an hour.

We spent the next 20 minutes watching a short film presentation in the visitor center auditorium. The film, *Rocky Mountain National Park: Spirit of the Mountains* (Radford, 2001), is played every 30 minutes for the rotating visitor population. The video contained several scenes that fit into our landscape-based communication rubric, and one scene emphasized climate change directly: "Since the end of the last ice age, about 10,000 years ago, the park's glaciers have gradually retreated warmed by the Earth's changing climate and melted back by the sun's persistent and powerful rays" (Radford, 2001). Another scene illustrated the connection to nature that the park provides: "The park's wild landscape contains some of the most spectacular scenery on Earth and provides an opportunity to restore our connections with the natural world" (Radford, 2001). Though not explicit, there was an attempt to provide a systems-based explanation of the interrelationship between human and ecological processes, but it was not connected to climate change: "While these high altitude specimens are extremely hardy, they are vulnerable to human trampling" (Radford, 2001).

Landscape-based Discourse and Public Engagement 71

The film also emphasized the importance of preserving the park for future generations, appealing to the visitors' family and civic values:

> Recognizing that lands are vital to the human spirit and diversity of life on Earth, the movement to save our national treasures in the name of future generations became symbolic of the democratic ideals of a growing nation...
> Today's travelers to Rocky Mountain National Park join in enduring human perception across time and place seeking to experience the same wildness as those who went before and those who will inevitably follow.
>
> (Radford, 2001)

After viewing the film we browsed the bookstore, and among a small collection of children's books we found *The Down-to-Earth Guide to Global Warming* by Laurie David and Cambria Gordon. This book, which discusses climate change at the third- to sixth-grade level, is divided into four parts: the science of global warming and an explanation of why it is happening, the effect on the earth's weather systems, the impact on plant and animal life, and simple things that children can do to help address the problem. We also found this book at another visitor center inside the park, and after our conversation with the first ranger, we were delighted to see that the NPS was promoting this book.

In our search for messages situated in the local landscape and targeting the wider visitor audience, we toured all the visitor centers and museums; none had brochures or exhibits specifically related to greenhouse gases or climate change in general. There was a series of interpretive posters about weather and quickly changing mountain climates in the museum but no mention of the impacts of the global climate change process we were curious about. This focus on science and ecosystem processes was reinforced as we coded the notes and recordings from our time at Rocky. The discourse centered on the general science of ecosystems, and although we saw these messages as an opportunity to include climate change science relevant to that particular landscape, it appeared that the park management was being strategically elusive and not using their federal status to invite a nonpoliticized, science-based conversation about climate change.

We concluded our visit by stopping at the visitor center at the north entrance to the park, where we spoke with two more rangers. The second ranger was much more willing to make an effort in responding to our request for climate change information. This ranger confirmed that they had no brochures, but he was optimistic that the park would offer new interpretive programs that could possibly cover climate change and supplied us with a phone number to check for program updates. This ranger was much more knowledgeable about climate change and appeared more comfortable sharing information; he entertained us with a 10-minute discussion of the shrinking polar ice caps.

We called later in the summer and found that the park had added a climate change–specific interpretive talk to their line-up of summer ranger-led programs. The program, Never Summer, Ever Summer, is offered every Friday (June 15–September 1) at 10 a.m. Participants are invited to "view the sun through a telescope and learn of potential effects of climate change on this park" (Rocky Mountain National Park, 2008). Through a deeper Web-based search, we found a June 16, 2008, report on climate change in the park. The report, *Climate Change in Rocky Mountain National Park: Preservation in the Face of Uncertainty* (NPS, 2008b), highlighted anticipated effects of climate change on the park's birds, mammals, lakes, streams, wetlands, and ecosystems and also addressed threats related to fire. All these information sources about climate change are invaluable, but when it comes to citizen engagement, these documents and presentations fall short because there is very little if any discussion of humans' role in climate change. For example, the *Climate Change in Rocky* report focused on climate change impacts on the landscape and management's response to such impacts. The report was concise and informative and promoted making management decisions related to climate change. There were three bullet points related to their education and interpretation strategy:

> This report will serve as a general outline of expected climate change impacts and collaboration opportunities.
> The information gained through the workshop will be presented to park staff during a one-day workshop and at the Park's 2008 Biennial Research Conference.
> The Continental Divide Research Learning Center will distribute this information in other formats and with other audiences as opportunities arise.
>
> (NPS, 2008b, p. 17)

We interpret these three bullet points as the starting point for Rocky Mountain National Park's development of a landscape-based public-engagement strategy. This report combines explicit examples of climate change impacts on the park, and now the challenge is to translate the scientific evidence into visitor-friendly messages. In the span of three months, we witnessed the early stages of this report's incorporation into three different interpretive programs. The first was a special ranger-led program, hosted in the evenings at the in-park campgrounds, titled The Pika—Ice Age Fortune-teller. This program described how the tundra Pika's evolutionary history is traced back to the ice age. The ranger explained that this species is signaling climate change today because of its rapid relocation to higher elevations. A second program, Climate Change at Rocky and Beyond, explored the causes and effects of climate change at Rocky. Again, the ranger explained how the landscape has changed, focusing on shrinking glaciers, reduced stream flows, and the influx of nonnative species and diseases. One of the major climate change culprits was emissions from vehicles, and using the park's shuttle bus was promoted as a simple solution for park visitors. In earlier visits, sustainability initiatives and green management efforts, such as the shuttle buses included in the Climate Friendly Park program, were not directly linked to public discourse about climate change at the park. A third program, Balancing Human Use and Preservation at Rocky Mountain National Park, explicitly used systems-based explanations to link human behaviors to environmental impacts, with appeals to enjoying the park as stewards of it.

Although the evidence during our May 2008 visit did not provide much beyond a children's book and promotional video, three months later we found numerous examples of landscape-based discourse in ranger-led programs and campfire talks. We can only speculate that this transformation was related to the June 16 report on climate change and that this evolution of climate change communication will continue at Rocky Mountain National Park. Park personnel have a rare opportunity to use its landscape as a climate change communication medium, reaching an average of nearly 3 million captive audience members annually (NPS, 2008a).

In April 2009 a special half-day climate change program was offered at Rocky Mountain National Park. The program was a combination of ranger-led talks and walks through three different landscapes affected by climate change, highlighting pine beetle infestations, the changing dynamics of a glacial lake, and the rapidly changing alpine tundra. First, we toured the most popular campground at Rocky Mountain National Park, Sprague Lake. Piles of "beetle kill" lined the gravel road along the perimeter of the campground. In the center of the facility was a makeshift amphitheater surrounded by hundreds of tree stumps. The clear cutting of lodgepole pines looked devastating, but the rangers explained that they are using this dramatic landscape change as an opportunity to talk to park visitors and campers about humans impacts on climate change —specifically, how the changing climate has made the destructive pine beetle more resilient whereas trees have become more vulnerable. With warmer winters, beetle larvae thrive into the springtime, but the pines need several days of deep, cold frost to kill the parasite. Without the freeze, the beetles are free to take over the nutrient systems of the trees. The rangers and park scientists explained the dynamic relationship of the pine beetle with this landscape. They also explained the need for management and intervention, despite the fact that actions such as clear-cutting are not typical Park Service duties. Many visitors are shocked to see a meadow of stumps, and this dramatic change has led to questions and discussions about the impacts of climate change at Rocky. This single example demonstrates that landscape-based discourse about climate change can provide park visitors with a systems-based explanation while illustrating the impacts of climate change on the immediate landscape. Through such conversations, and because of the Park Service's federal status, it has the power to change the conversation about climate change—from ideological diatribes to proactive political discussions about managing for change and responding to the immediate evidence of a rapidly warming planet.

Yellowstone National Park
Yellowstone National Park (Yellowstone) welcomes more than 3 million visitors per year and has seen more than 140 million visitors since the park opened in 1872 (NPS, 2008a). Yellowstone consists of over 2.21 million acres at the intersection of Wyoming, Montana, and Idaho. As in Rocky Mountain, our time at Yellowstone was spent talking to rangers, discovering exhibits, and touring visitor centers in search of climate change messages. We arrived at the information station and bookstore located in west Yellowstone as they were opening. After wandering around the station with no luck finding information on climate change, we approached the closest ranger with our questions. The ranger politely stated that she was unaware of any publications, programs, or information that the park had about climate change. We pursued our quest by asking if she knew anything about the topic that she could share with us, and she quickly responded that she had not received training on the issue.

We then headed to the Junior Ranger Station. We entered the station as a park ranger was playing an environmental education game with a few attentive children. We approached the front desk to request information from the ranger on duty about climate change. Though the ranger seemed very interested and upbeat about the subject, she stated that this station was intended for young children and did not cover such complex topics. She encouraged us to look at the other visitor centers.

We then stopped at the Old Faithful Visitor Center. Along with many other visitors, we swarmed straight to the visitor center door to find out when Old Faithful was predicted to erupt again. We learned that the show would not begin for another 30 minutes, so we inquired about climate change impacts instead. Glancing up at the line of tourists with questions, the ranger excused himself to help us. He immediately stated that the park did not carry any information specifically on climate change or global warming; the Old Faithful Visitor Center predominately focused on wildlife and matters concerning geysers. Busy with other tourists, he moved to the next visitor in line. With minutes to spare we headed out under our umbrella with hundreds of other spectators; we waited in anticipation for Old Faithful. Following the brief minutes of the geyser's eruption, we overheard a couple comparing the size of that day's eruption to one they had seen on a past trip to Yellowstone. Other tourists were eavesdropping on their conversation, and a young woman offered that she had heard a drought was making Old Faithful's eruptions smaller. Could these park visitors have their own systems-based explanations for changes in the Yellowstone ecosystems?

We then moved on to the Canyon Village Visitor Education Center, hoping that because it was an education center it would offer information about the impacts of climate change on this landscape. The first ranger thought that the park may bring in outside experts to have informative talks but was not sure whether this had ever happened or when the next presentation might be. His colleague excitedly rushed over after hearing our questions and referred us to a book that was available for purchase in the bookstore: *Yellowstone Resources & Issues* (NPS, 2008d). This book is an annual compendium of information about Yellowstone National Park and contains eight pages on climate change among the 200 pages covering a multitude of issues. The bookstore sales-

Landscape-based Discourse and Public Engagement 77

clerk mentioned that it was the best and most informative book that they sold in the store. He also made clear that it was not a bestseller. The book has a scientific format with more text than photos; visitors probably opt for the glossy pages and panorama pictures in the Yellowstone coffee-table book. We skimmed though 171 pages before coming across four that focused on climate change. Climate change in Yellowstone does not appear to be a pressing issue thus hidden among layers of other problems and park history.

Many of the communication themes found at other parks, which we discuss later, were missing at Yellowstone owing to the lack of climate change information. The dominant messages available to visitors fit under the categories of sustainability initiatives and green management efforts. In 2008 Yellowstone founded Yellowstone Environmental Stewardship (YES!) in order to meet sustainability goals by 2016. The program aims to increase operational efficiencies and to reduce the park's ecological footprint. Yellowstone has also applied to be in the Climate Friendly Park program but has yet to meet any of the requirements. If Yellowstone is able to link green management efforts to greenhouse gas emissions, it may empower park visitors to take action and reduce their contribution to climate change impacts.

Our final stop was at the Mammoth Hot Springs Visitor Center. The ranger there confirmed that Yellowstone, as of June 2008, had no brochures or literature that visitors could pick up to read or take home. Before we left he mentioned that *Yellowstone Resources & Issues*, the book we had examined at the last visitor center, might include a few pages on climate change and might be worth looking at. Although the book does include detailed information on climate change and its impacts on local landscapes, it is available only to those willing to purchase it. It appears that Yellowstone is not yet prepared or equipped to share climate change messages with visitors.

Following a fruitless on-site visit, we took a virtual expedition and did not find one statement about climate change on Yellowstone's website.

Again, playing the role of the typical tourist, we observed numerous opportunities for Yellowstone to engage park-goers in a public conversation about the nuances of climate change in the west, especially on this treasured landscape. Maybe our assessment is premature; perhaps, as at Rocky Mountain National Park, there will be a groundswell of interpretive programs and literature in the near future. However, at this rate we question the agency's ability to be a leader on the issue, despite director John Jarvis's commitment: "One of the most precious values of the national parks is their ability to teach us about ourselves and how we relate to the natural world. This important role may prove invaluable in the near future as we strive to understand and adapt to a changing climate" (NPS, 2012b).

Glacier National Park
In no park are the effects of global climate change more apparent than in Glacier National Park (Glacier). The park was established in 1911 as the 10th national park, its boundaries securing more than 1 billion acres of land. The park's name comes from the remnants of the ice ages that occurred 10,000 years ago. Though the name is still relevant, some wonder how long it will remain so, for the receding glaciers, visible from the roadside, provide evidence of a quickly changing climate.

As we approached the entrance, we predicted that Glacier National Park would provide the most extensive landscape-based discourse about climate change. The park upheld our prediction as we entered and received the official Glacier Visitor Guide. On the bottom of the cover was a brief preview of the climate change story featured on page 8. It became apparent that Glacier National Park is seizing the opportunity to teach visitors about the present and future of climate change:

> [National Parks] help us to understand the extent of climate change, how to mitigate its effects, and how to protect natural and cultural treasure for the enjoyment of generations to come ... [The] changing environment provides a powerful example of

Landscape-based Discourse and Public Engagement 79

what could be lost without global action to reduce greenhouse gas emissions ... Glacier's diverse landscape ... show[s] how climate change affects an intact ecosystem.

(NPS, 2008c)

The article recognizes the extensive impacts climate change is causing in Glacier. It provides a systems-based explanation of social-ecological changes that are taking place on the landscapes that tourists treasure. Another section presents a shocking visual appeal to demonstrate the disappearing glaciers. A photo of Shepard Glacier in 1913 is compared with one taken in 2005. The photo from 2005 shows significant glacier retreat; only a fraction of the glacier remains. Despite some uncertainty in the exact number of years, the article estimates that in 1850 there were 150 glaciers in the park 50 in 1968, and now only 26 glaciers. Scientists predict that the remnants of these glaciers will be completely eliminated by 2030 (NPS, 2008c). To some this may appear a hopeless fight, but the park has taken responsibility: "Glacier National Park strives to be a leader in educating park visitors about climate change" (NPS, 2008c).

By the time we made it to the Apgar Visitor Center in west Glacier, it was raining and a layer of fog was hanging from the sky. We pulled on our raincoats and dashed into the building. As we looked around, we saw about a dozen tourists attempting to dry off as they meandered around books and displays.

We waited until a hiker finished reporting a bear and cub sighting off a nearby trail. A ranger then attended to our climate change inquiry. He told us that the park currently had no educational or interpretive programs on climate change but that he hoped climate change would be discussed in the next ranger training. Two other rangers mentioned the name of a United States Geological Survey (USGS) climate change scientist, and they noted that this scientist occasionally gave talks to visitors and in the adjacent community.

Nearby, at an environmental education classroom, another ranger explained that Glacier is a unique place because climate change can be measured and monitored within the park's boundaries. This ranger noted that the on-staff climate scientist measures the glaciers regularly to identify how much Glacier National Park is being affected by climate change.

While perusing the bookstore, as many tourists do, we found *The Atlas of Climate Change: Mapping The World's Greatest Challenge* by Kirstin Dow and Thomas E. Downing. This 112-page book provides the latest evidence of climate change as well as a portrayal of the past, present, and possible future of the earth. The book covers a wide range of topics, including warning signs, future scenarios, vulnerable populations, health impacts, renewable energy, emissions reduction, and personal and public action. We asked how often the book is sold, but the salesclerk did not know.

The visitor center had one climate change brochure that visitors could pick up and take home. The brochure targeted climate change in Glacier National Park. It emphasized the importance and meaning tourists place on glaciers within the park: "For many people, the glaciers are a key reason the park holds special significance and are a feature they expect to see when they visit" (NPS, 2007b). The park takes advantage of the visitor's connection with the landscape to show that it has more than one meaning: "Mountain glaciers are more than just scenery, they are an integral part of the ecosystem, providing cold water to mountain and downstream environments" (NPS, 2007). The brochure stresses the importance of individuals' actions to save the landscapes they feel an attachment to, focusing on the connection tourists have with nature and the positive changes they can make: "Any actions or choices that can result in a reduction of these emissions will put us on a more sustainable path toward stewardship of the resources we are charged to protect" (NPS, 2007). The brochure ends with examples of efforts to make Glacier National Park more energy efficient and the management's dedication to raising awareness about climate change. Overall, the brochure was used as an opportunity to inform visitors of climate change impacts on the glacial landscapes through place-based messages.

The next day we headed to the Saint Mary Visitor Center in east Glacier, but upon arrival we discovered that it was closed for renovations. We continued along the National Historic Landmark Going-to-the-Sun Road, an engineering project that took the better part of 11 years to complete. On our way out of the park, we came across an interpretive sign on the side of the road entitled *Going, Going, Gone* (figure 1). The roadside display asks the reader to look at and listen to the landscape:

> Do you see a glacier? ... Visitors today see only 25% of the ice that existed in 1850, and projections are that the park's glaciers will be gone by 2030. Of the estimated 150 glaciers present in 1850, approximately 26 remain.... The park may look different on your next visit, because the recession of glaciers like Blackfoot and Jackson affects the entire ecosystem in many ways.

As we did for the other parks, we followed up Glacier's climate change programs and communication messages by browsing its webpage. Glacier National Park offered a special ranger-led program in mid-June called Goodbye to the Glaciers. The program is offered weekdays, every hour between 11 in the morning and 4 in the afternoon. Though the program is only 15 minutes long, it provides the casual park visitor the opportunity to see the impacts of global climate change on the glacial landscape.

Discussion

Our analysis has shown that in some ways parks are using their landscape to engage visitors with climate change. Protected areas, as a material medium, offer the potential to engage millions of visitors in new understandings and new conversations about climate change.

Figure 1. Going, Going, Gone (wayside interpretive sign at Glacier National Park).

Source. Photograph by Sarah Schweizer, May 2008.

Taking into consideration the visitors' attachment to the landscape, their sense of stewardship for the parks, and experiential learning opportunities provides a platform to increase awareness of climate change–related issues and promote political action to mitigate further impacts.

Landscape-based discourse requires using place as a medium and connecting that place to emotional, rhetorical, and political messages about the impacts of climate change. Because many people need to see the effects of climate change before they can believe it is real, the parks can serve as a national ambassador, facilitating nuanced conversations about climate change and engaging citizens in political, social, and environmental activism on related issues.

While we were touring the parks, playing the role of typical visitors, we were simultaneously involved in a national-level NPS research project. In the past year, we have noticed a gap between the Washington-level effort to communicate climate change and the tourist experience at Rocky, Yellowstone, and Glacier. Our Park Service partners provided us with a service-wide brochure, explaining the impacts of climate change on all of the parks:

> For our national parks to thrive and for us to continue enjoying them, it seems appropriate now to do what we can to reduce climate change impacts and adapt to their consequence. Fortunately, we now have the tools, knowledge, and ingenuity to better understand these changes and make informed choices for coping with them.
>
> (NPS, 2008e)

The brochure informs visitors about simple actions they can take, such as using energy-efficient appliances and lightbulbs, unplugging electronic devices, and using public transportation. The NPS also included a list of websites at which visitors can learn more about climate change science, behavior changes, and climate change policy (e.g., the Intergovernmental Panel on Climate Change, the Arctic Climate Impact Assessment, Understanding and Responding to Climate Change, the EPA's Global Warming-Actions, and NPS/NASA Earth-to-Sky Interpretive Training).

In addition, we have seen several brochures created by the USGS with the Climate Change in Mountain Ecosystems (CCME) program in Glacier National Park. The USGS has used repeat photography to assess and communicate the effects of climate change on landscapes. These features are also available by virtual tour. The USGS has argued that landscape-change photography provides "powerful images, with their inherent ease of interpretation, [that] have become icons of climate change" (USGS, 2007, p. 2).

The CCME website not only hosts the repeat photography demonstration but also provides a vast number of resources—posters, movies, photos, podcasts, PowerPoint presentations, models, publications, and projects—meeting the need of almost any audience. The information is closely tied to the visitors' emotional connection with places they visit and provides a virtual experience that in many ways is a contemporary example of new technology's role in landscape-based discourse about climate change.

The NPS, as a bipartisan federal agency, has an opportunity to take advantage of the potential for facilitating landscape-based climate change engagement with park visitors and stakeholders. Climate change discourse would be vastly improved even by simply ensuring that these resources were being shared with the everyday tourist as opposed to being handed out at Washington-level meetings. We conclude by suggesting if the NPS wants to be a leader in communicating about climate change and citizen engagement, it should develop and provide landscape-based messages for park visitors, linking the valued places visitors have come to experience to the issue of climate change. To make a significant impact on visitors' perceptions and actions related to climate change, park authorities must find a way to connect people to tangible and accessible issues in the parks (Thomashow, 2002).

Scientists, communicators, and stewards of public lands should focus on presenting local climate change impacts that are occurring right now, forcibly making the issue one that is current and salient to community members and decision makers. Providing relevant, contextual examples will encourage and influence individuals to make positive changes in order to combat climate change on a local, regional, national, and global scale. Communicators should avoid crafting messages that highlight risks and uncertainty, focusing instead on creating messages that target personal points of reference and ties to landscapes (O'Neill & Nicholson-Cole, 2009). Past research has shown that increased risk perceptions only lead to increased protective behaviors and do not motivate individuals toward effective behavior change (Leiserowitz, 2005; Moser & Dilling, 2004; Plotkin, 2004). Aust and Zillmann (1996) have shown that emotionally resonant messages are more memorable and accessible when an individual needs to make a decision or articulate a judgment about an issue.

Global climate change has been eloquently connected to polar ice caps and polar bears, making it appear a distant problem that may affect only people and animals far away. In Leiserowitz's (2007) research, survey respondents were able to list only extreme examples of the impacts climate change may have, such as the melting of polar ice caps, as opposed to the more common daily consequences (e.g., cherry blossoms blooming two weeks earlier). Individuals rarely perceive climate change as a local issue or as related to places they value (Bostrom & Lashof, 2007). Furthermore, Leiserowitz (2007) found that only a minority of respondents were even concerned about effects on themselves, their families, or their local communities. Several studies also discovered that local threats were generally perceived as more salient and of greater urgency and importance than global or distant problems (Leiserowitz, 2007; O'Neill & Nicholson-Cole, 2009). Americans seem to accept climate change or global warming as a real phenomenon, but most do not seem to have a great deal of concern about it (National Science Foundation, 2006). Therefore although Americans believe in the power of individual action and the ability of regular people to make a difference, they have failed to connect the impacts of their individual actions on the larger system of global climate change.

We studied three popular parks in the intermountain west of the United States, and it is possible that other national park regions and protected landscapes, such as United States Forest Service territory, state parks, and municipal parks, may have more developed communication and educational programs on climate change. Each landscape will experience diverse and pertinent consequences from climate change that may lead to unique, landscape-based discourse. We encourage future researchers to analyze landscapes and climate change engagement in other regions of the world. Protected areas, especially federally managed lands, represent an opportunity to take advantage of their unique political status as well as their publicly treasured landscapes as a forum for renovating public understanding about climate change. Communicating climate change in parks moves the issue into people's backyards. Messages about climate change are often politically divisive and controversial, but showcasing place-based impacts provides tangible examples in the places visitors are already connected to and concerned about. *Backpacker* magazine (2007) published *The Global Warming Issue*, which highlighted the devastation America's parks will see in 50 years owing to climate change. The national parks are in a rare position to use such media attention to deepen the public's understanding of climate change as it impacts the beautiful and grand landscapes of the United States.

Park authorities can help facilitate and inspire citizens to become publicly engaged by creating messages that address the connection between individual action with how visitors think and feel about climate change. Parks have the power to change the political communication of climate change from a muddle of technically focused and highly politicized disputes to a suite of conversations about the places people treasure. Engaging park visitors in this way sends them home with a unique experience that they can begin to translate into their daily decision making and ultimately the policies and political platforms they support. The NPS and other U.S. land-management agencies should examine their roles and the power that they have to foster a new and productive discourse on climate change through public engagement at their sites. Such agencies are public servants, now serving as eyewitnesses to the effects climate change has on national treasures. They have an obligation to engage the American public and other visitors in candid conversations about climate change and its impacts on public lands. The NPS needs to vigorously exploit its opportunities to encourage more widespread public participation in climate change politics, which is critical to shifting the political tide on climate change policy in the United States.

CHAPTER 4

THE VISUAL RHETORIC OF CLIMATE CHANGE DOCUMENTARY

Helen Hughes

INTRODUCTION

In his book *Natural Visions: The Power of Images in American Environmental Reform* (2005), Finnis Dunaway argued that "the history of environmental reform is more than the passage of a series of laws; it is also the story of images representing and defining the natural world, of the camera shaping politics and public attitudes" (p. xvi). In the course of his discussion on the use of photography and film in environmental debates and campaigns from 1900 to 1970, covering the progressive era, the New Deal Era, and the 1960s, Dunaway described a development from the sublime and the romantic to the violent and destructive, to "wonder in the small" (Dunaway, 2005, p. xix). In other words, during this period, the images that have represented the environment people see as natural have moved toward and then away from the long shot, the panorama, or the bird's-eye view.

In contemporary environmental documentary filmmaking, high-angle extreme long shots, including aerial shots and space photography, continue to reinforce the role of the camera and its technological extension, remote sensing, in "shaping politics and public attitudes" (Dunaway, 2005, p. xvi). Part of the iconography of the "ecodoc" is the aerial view, which reveals in a spectacular way the impact human society makes on the landscape. This chapter seeks to contribute to the growing literature on the use of images in communicating climate change by examining the role of aerial cinematography in contemporary climate change documentaries.

The focus on the aerial shot is on a contested category of image making that since the Second World War has played a profound role in environmental planning. In his foreword to E. A. Gutkind's *Our World from the Air* (1952), G. P. Gooch underlined the significance of this perspective to the new generation that will be responsible for designing the fabric of the human environment for the second half of the century:

> That we are living in a new age we discovered to our cost in the Second World War. But the discovery of the art of flight enables us to do a hundred things in addition to bombing our enemies and racing about the world. It allows us, for instance, to survey every portion of the earth's surface from a new angle of vision, to supplement the detailed though limited approach of the earth-bound observer by a synthetic perception which reveals familiar objects in a fresh light. Perspective is the key to understanding, and the whole is greater than the part. There is a new fruitful technique for the geologist, the archaeologist, the town-planner, the sociologist.
>
> (Gutkind, 1952, n.p.)

What is contested in this vision, as I discuss in what follows, is the belief in the aerial shot as "the truer view of the world" (Gutkind, 1952, n.p.)

The Visual Rhetoric of Climate Change Documentary 89

Two quite distinct and opposing directions can be discerned in attitudes toward aerial images since the 1950s. The first is the pessimistic concern that still- and moving-image technologies, integral to the predominantly visual culture of modernity and particularly significant in the development of remote-control surveillance and weaponry, have themselves contributed to the distortion of habitable space through their contribution to military, urban, and industrial planning. The second, however, represented by such publications as Yann Arthus-Bertrand's *The Earth from the Air* (2005), takes the more optimistic view that aerial images, particularly of the whole planet, can not only show what has gone wrong but also inspire people to love and take care of the environment. In this latter view, the aerial shot provides visible evidence for the considerable effects of centuries of agriculture and industry on the environment, but it also opens up a dialogue between different political perspectives on the landscape as observed and analyzed by scientists, framed by policy makers, and experienced in everyday life.

This chapter offers an analysis of the integration of the aerial perspective in three climate change documentaries made in the United States in the first decade of the 21st century. In a context in which the U.S. federal government appears to have failed to respond politically to international attempts to create a climate change policy, these documentaries in their different ways seek to reconcile various perspectives around the single point that the community of scientists and activists, new and old, are united (in their diversity) through their belief that the threat is real and that political action can make a difference. Aerial cinematography contributes to this strategy of communication through its metaphorical expression of the unity of the planet. At the same time, however, the aerial shot poses a number of problems. Although the technology to make aerial shots developed rapidly during the 1990s, it still represents a form of filmmaking that is economically and environmentally expensive and strongly associated with military power (Wiener, 2000). Its use therefore needs to be strategic for several practical and symbolic reasons.

Understanding the significance of the integration of aerial images requires some analysis of the aesthetic debate regarding the appropriate way to represent the relationship between humankind and nature. This chapter begins by sketching some of the elements of this debate as it has played out in the context of environmental filmmaking. In addition, Julie Doyle has drawn attention to the problems regarding the use of photography per se in campaigning on climate change, arguing that global warming "prompted a crisis of representation, and thus communication for environmental groups" (Doyle, 2009, p. 280). She pointed out that the photograph is able to show only what has happened to the landscape and is not in fact capable of indicating future threats. Doyle fears that photography has a paralyzing rather than a galvanizing effect because it indicates that the damage is already done.

The question about the use of aerial photography in filmmaking is not the same as this argument, but a related critique of its effects can be made. Aerial photography is a point of view that favors large-scale dramatic impacts carrying with them the implication that individual human perspectives in the here and now are irrelevant in the face of such catastrophe. These criticisms of photography as a campaigning tool lead to the conclusion that the use of visual images has negative, depressing effects and does not motivate the public as part of a public-information campaign. The goal must be, therefore, to find policies to exploit what is undoubtedly effective about visual campaigning while finding ways to counteract its potential negative effects.

The third part of this chapter examines the ways in which contemporary climate change documentaries released between 2005 and 2010 have responded to these critiques by embedding aerial shots into human-scale stories, incorporating both the spectacular power of the aerial shot and the personal relevance of the individual and community perspective. Though the spectacular nature of the overview remains evident, attracting attention and provoking emotional responses, contemporary uses of aerial images in environmental documentaries on climate change appear to be about preventing the power of the aerial perspective from displacing policy issues entirely onto long-term and large-scale projects by using aerial photographs and aerial shots as artifacts, products of science and art, and articulating alongside them, on a human scale, campaigns concerning the day-to-day realities of the present.

In conclusion, I argue that the aerial view, embedded as it is in the research and evidential basis for environmental change more generally and for climate change in particular, is an important resource for attracting audiences and communicating impacts through the documentary film form. What is required for the use of photography in climate change campaigning is an understanding of film production as part of a dynamic process in which individuals and communities take control of the medium as well as of the message that it conveys. Contemporary documentary filmmaking, concerned as it is with creating empowering relationships among filmmakers, communities, and audiences, offers a representational resource that goes some way toward achieving this goal.

Ambivalent Perspectives

Like Dunaway, Scott MacDonald, in his *Garden in the Machine: A Field Guide to Independent Films about Place* (2001), pointed to an oscillation between the panoramic shot and the close-up in experimental film, a form of filmmaking practice that in the United States from the 1970s to the late 1990s was deeply concerned with the relationship between human society and the environment (S. MacDonald, 2001). In MacDonald's account, filmmakers of the 1970s interpreted the panoramic as the expression of the age of empire at the end of the 19th and the beginning of the 20th century, and turned to the close-up as a radical alternative. This shift became part of an attempt by experimental artists, such as Carolee Schneemann and Stan Brackhage in films such as *Fuses* (Schneemann, 1964–1966) and *The Garden of Earthly Delights*

(Brackhage, 1981) to fuse the sense of self with the sense of landscape. It is an approach that resonates with sociologist Henri Lefebvre's attempts to recapture alienated space from the distorting eye of photographic technology:

> Wherever there is illusion, the optical and visual world plays an integral and integrative, active and passive, part in it. It fetishizes abstraction and imposes it as the norm. It detaches the pure form from its impure content—from lived time, everyday time, and from bodies with their opacity and solidity, their warmth, their life and their death. After its fashion, the image kills.
> (Lefebvre, 1974/1991, pp. 96–97)

This passage from *The Production of Space* sets out the philosophical problems with the visual portrayal of the relationship between humankind and the environment for poststructuralist and then phenomenological analyses of social activity. Lefebvre's discussion insists that behind all definitions of space of whatever kind there lie hidden the economic and political positions of power. Images do indeed represent and define the natural world; they do shape politics and public attitudes, as Dunaway argued, and in so doing they distort reality and prevent people from experiencing space as place in their own way, in their own space and time. However, Lefebvre did not dismiss photography entirely as a medium that has the potential to reveal social space. "Occasionally," he wrote, "an artist's tenderness or cruelty transgresses the limits of the image. Something else altogether may then emerge, a truth and a reality answering to criteria quite different from those of exactitude, clarity, readability and plasticity" (Lefebvre, 1974/1991, pp. 96–97).

MacDonald's account of experimentation with the camera in the 1980s and 1990s reads as an attempt on the part of filmmakers to rescue the media of film and photography and to deal with this argument by creating a cinematography capable of expressing the fullness of everyday experience. (see also Willoquet-Maricondi, 2010) In so doing, the art filmmaker becomes a political activist through the attempt to alter the cate-

The Visual Rhetoric of Climate Change Documentary

gory of the visible, or in the way described by Jacques Rancière, "the distribution of the sensible" (Ranciere, 2004, p. 12). Rancière defined this as "the system of self-evident facts of sense perception that simultaneously discloses the existence of something in common and the delimitations that define the respective parts and positions within it" (p. 12). Shared visible space is what is at stake in representations of the earth and the debates about its future.

Rose Lowder's "ecological cinema" can be seen in particular as a concerted attempt to find a consistently more sustainable way of making small and intimate films in the 1990s, attending not only to the aesthetics and sociology of the image but also to the question of the materiality, particularly the wastefulness of most filmmaking processes, taking the debate yet further. Lowder's work anticipates the turn events took in the first decade of the 21st century. At the end of the 1990s, however, MacDonald also found a renewed interest among independent filmmakers in the long shot and in the aerial shot. The industrial sublime took over from the naturally spectacular images of mountain ranges and coastal landscapes, visualizing and mythologizing instead the large-scale impacts of technology on the environment. Bruce Connor's *Crossroads*, for example, is a montage of nuclear explosions filmed from many different angles. Following the tradition of the compilation film, Connor took these shared images—"the raw material of history" (Barnouw, 1993, p. 198)—and demonstrated their aesthetic rather than their historical qualities. In so doing, Connor made visible the links between aesthetic history and environmental devastation. MacDonald described the difficulty in acknowledging this link and thus the pause for thought about the use of structural form and the aerial perspective for understanding history and the formation of policy on the shape of the future:

> Since the detonations are visually arresting, even gorgeous to look at, and since this aesthetic impact is enhanced by the frequent slow motion and by the Reilly score, one is tempted to fall into a meditative sensibility.... And yet our consciousness of the *func-*

tion of this visually sublime process—its function in history, its implicit goals—renders any such meditation conflicted at best.
(S. MacDonald, 2001, pp. 323–324)

The understanding of Connor's work as ecological cinema includes the representation of the impact of technology on the environment. Werner Herzog's *Lessons of Darkness* (1992) applies a similar aesthetic to contemporary destructive events, using CNN news footage and filming the repair of oil wells after the first Gulf War from a helicopter. For Herzog the physical challenges in obtaining the footage represent a considerable part of the art of filmmaking. Thus when he described the skill of the pilot, he was indicating a form of heroism in managing to get the images he wanted: "For a project like this a good pilot is as important as a cinematographer. He had to understand the terrain and air flows around the burning oil wells and establish a pattern of flight to facilitate a sequence of travelling shots.… If he had flown into an area where the heat might suddenly be blown towards it, the helicopter would immediately explode" (Herzog, 2002, p. 247). In this case MacDonald's analysis of the film accepts the sublime dimension of the viewing experience as clearly expressive of major political and environmental disaster (despite the fact that Herzog did not identify the images as from Kuwait), reading Herzog's subtitle to the film, *Satan's National Park*, as a critique of the expansionist policy pursued through the war:

> The sublimity is a function of the immensity of this disaster, of the size of the fires and the towering billows of black smoke; and it is powerfully confirmed both by Herzog's use of helicopter shots to survey the oil fields and by his accompanying the imagery with a sound track composed of excerpts from classic works that suggest the epic scope and historic significance of what we're seeing.
> (S. MacDonald, 2001, p. 326)

Herzog's relationship with the Romantic sublime is complex and connected with his own search for "ecstatic" images that are missing but in his opinion needed to express the contemporary world. The aerial

The Visual Rhetoric of Climate Change Documentary 95

perspective is accompanied by Herzog's own soft voice, which expresses tenderness combined with cruelty in a way that Lefebvre argues "transgresses the limits of the image" (Lefebvre, 1974/1991, p. 96). Understanding the past from the perspective of the future, placing the images in a fictional context Herzog called it a science fiction film, "a requiem for a planet we ourselves have destroyed" (Herzog, 2002, p. 249). Not only are different things visible from the planetary perspective, but the balance of things is changed so that values also change. Including two interviews with victims of torture, Herzog emphasized the cruel, inhuman nature of a planetary scale that renders the individual mute. The politics of the image relates to its configuration of the sensible, in which the individual is enveloped in a much larger disaster.

In their recent book *Photography and Flight*, on the history of the suturing of these two technologies, Cosgrove and Fox provided another analysis of the use of the "God's-eye view" in the extraordinary development of aerial photography and remote sensing. Beginning with the earliest map known, Cosgrove and Fox understand the quest for the aerial view as an integral part of human nature.

> To view the world from above seems to be an innate human ability that is activated when we are very young. From the moment a child first picks up an object and tries to turn it in his hands, it begins to develop the skills to rotate first the micro-environment—things that are smaller than the body—and, with time, the macro-environment. The former allows the child to grasp, and thus to learn by feel and touch, those objects that are smaller than the self, and to understand and remember them as discrete wholes with a gestalt that is greater than its parts.... With age, that ability extends to things larger than our own bodies, to the macro-environment, and we learn effortlessly to view space from a myriad of imagined view-points. We transfer that early ability to feel an object in our hands to seeing from different angles the room in which we sit, the house in which we live, our neighbourhood and eventually anywhere we go and even places we merely glimpse in a picture or conjure out of listening, reading or imagining. This

capacity to picture places might be called the "geographical imagination," and it finds its most immediate graphic imagination in maps, plans and architectural drawings. It is one of our most basic methods of transforming space into place—that is, into known and meaningful locations and environments. It is also a fundamental navigational skill essential to survival.

(Cosgrove & Fox, 2010, pp. 10–11)

The technical developments and the practical and aesthetic uses to which the images are put start from the use of early photographic equipment to take pictures from a balloon in 1860. The vehicles used to take the camera up off the earth have always been varied and imaginative, from kites to miniature remote-control helicopter cameras. In introducing his survey of modern companies specializing in aerial photography in 2000, for example, David Wiener described a picture taken from a "pigeon camera" patented in 1903 that "shows a castle far below, nestled in rolling countryside. Along the right-hand edge, the sunlit feathers of the pigeon's wing can be clearly seen" (Wiener, 2000, p. 92).

This peaceful image is not, however, typical for the development of aerial photography and cinematography, which is closely linked to the development of surveillance during the First and Second World Wars and subsequently to the extension of surveillance to satellite images during the Cold War. The story ends (currently) with the integration in the 21st century of extraordinary quantities of data with remotely sensed images for public use on Google Earth. By the time the 21st century began, photography had ceded its primary position and become "an increasing part of a larger category of aerial images, at least in terms of most military and scientific applications" (Cosgrove & Fox, 2010, p. 70). This larger category includes space photography and satellite imaging, which involve significantly higher altitudes and mean that the technology is no longer attached directly to the human eye but operates remotely, sending images using first analog and then digital technology.

The Visual Rhetoric of Climate Change Documentary   97

Cosgrove and Fox explained how Evelyn Pruitt coined the phrase *remote sensing* to describe a whole new category of spatial documentation differentiating between aerial photography and satellite imaging. These categories have been blurred, however, by the development of remotely controlled helicopters and ever more sophisticated crewed helicopters to capture images at close range as well as from great heights. *Remote sensing* has consequently come to describe any image taken from a point above the surface of the earth, making contact with the earth itself into a category for image making.

Examples of the recently developed practical uses of remote sensing are significant for the discussion of the images used in arguments about climate change in environmental documentaries. Having been developed first as part of the war machine, the uses of aerial photography have now become critical in the understanding of the earth as an integrated ecological system. In 1962 the Canadian Department of Agriculture developed a computer-based system to correlate data on land usage in rural parts of the country. Starting in 1964 weather satellites were developed to provide information on atmospheric phenomena every 24 hours. From 1972 the National Aeronautics and Space Administration (NASA) began a geological survey of the United States using remote sensing for capturing images of the ground. In 1999 NASA developed the Earth Observing System (EOS) to examine environmental factors, such as air temperature, humidity, and aerosol concentrations, from the ground up to the stratosphere. With NASA overwhelmed by the amounts of data produced, Google Earth, which had developed applications after buying Keyhole, Inc., superimposed data on 3-D global maps in iEarth (Cosgrove & Fox, 2010; Jasanoff, 2004).

At the same time, however, the military and technological history of aerial photography also bears witness to ambivalence about the power of its perspective, as documentary filmmaker Harun Farocki, for example, has explored in films such as *Bilder der Welt und Inschrift des Krieges* (*Images of the World and the Inscription of War*, 1988) as well as in many installations. Cosgrove and Fox recounted the utopian impulses that derived from the valorization of the airplane during World War II:

> Political scientists and public intellectuals such as the poet Archibald MacLeish wrote that World War II was the "airman's war." The airman, he argued, gained a unique perspective of the earth to see things "as they truly are." Removed from the petty squabbles and concerns of daily life on the surface, the airman could dream of utopian futures: "The airman's earth, if free men make it, will be truly round: a globe in practice, not in theory."
> (Cosgrove & Fox, 2010, p. 58)

The transfer of technology from wartime to peacetime uses is thus at the center of the critiques of postwar America and Europe, critiques that point to the ways in which the rigid patterns of town planning and the large-scale impacts of land use in farming and mining are enabled in part through the aerial perspective. Cosgrove and Fox pointed to another twist in the tale in an account of the sources used for the Sierra Club's famous first coffee-table book, published in 1960 to support the cause of wilderness conservation, *This is the American Earth*. William Garnett's aerial images of housing development in postwar Los Angeles were originally taken to become part of the volume cited in E. A. Gutkind's introduction to *Our World from the Air* (1952), arguing for a more qualitative approach to town planning. In a pattern repeated across the world in response to postwar town planning, the images were used by Ansel Adams and Nancy Newhall simply to demonstrate the destruction of the environment (Cosgrove & Fox, 2010, p. 65).

Although Cosgrove and Fox argued that the "American environmental movement ... has ever since relied heavily on aerial photography to help make its case" (Cosgrove & Fox, 2010, p. 63), the use of the image is never straightforward. Aside from the military past there is the environmental cost of aerial photography itself. Although the economic costs of aerial photography may have come down, it is still more expensive than photography from the ground, and though it may be possible to

The Visual Rhetoric of Climate Change Documentary 99

use remote-controlled helicopter cameras or lightweight aircraft to gain height (Wiener, 2000), aerial photography is still not a particularly green form of filmmaking. Wiener's account of the exploits of aerial photographers in the making of fiction films mentions safety issues and noise pollution but not air pollution. He quoted a pilot, Al Cerullo, who

> has been flying movie helicopters around New York City for about 25 years; his credits include *Klute, King Kong,* and *Superman.* He says that New Yorkers have gotten more and more sensitive to the noise created by constant fly-bys. "There is a coalition in New York City [that works] against helicopters" involved in aerial cinematography, he reveals.
>
> (Wiener, 2000, p. 112)

Both its prevalence and its ambivalence point to aerial photography and cinematography as a contested category.

THE AERIAL PERSPECTIVE IN CLIMATE CHANGE DOCUMENTARIES

The aerial perspective can be found in the use of aerial photography, aerial cinematography, space photography, and satellite images in three U.S. climate change documentaries: *An Inconvenient Truth* (David & Guggenheim, 2006), *Everything's Cool* (Gold & Helfand, 2007), and *The 11th Hour* (Conners Petersen & Conners, 2007). These films do not contain an enormous amount of aerial footage in terms of screen time. In *An Inconvenient Truth* there are approximately 4 minutes of aerial and space images used in the film, each shot between 1 and 35 seconds long, with the mode at 3 seconds. In *The 11th Hour* the total is about 6 minutes, shots from 1 to 12 seconds, with the mode at 2 seconds, and in *Everything's Cool* only 13 seconds of aerial images are used.

What is significant about the documentaries is their realignment of the aerial perspective around a new problem that takes account of metaphorical and physical uses of the "overview" in climate change research. In a sense, the very issue of climate change alters the discourse around the use of perspective to represent the relationship between human beings and the environment. As discussed, destructive aspects of environmental planning become visible—perhaps in a distorted way—through the use of the aerial perspective. Housing arranged in grids, intensive farming that alters vast landscapes through its monocultures, opencast mining, industrial complexes, oil rigs and refineries, complex road networks, megacities, and urban sprawl are all impacts on the earth that are highly visible from the air. These can all be related through argumentation to climate change, but the phenomenon itself is not so easily made visible, although it is the warming of the atmosphere—the air—that is in question. Smoke and haze, particularly rising from cities in hot weather, and the visibility (once sufficient height has been achieved) of the earth's atmosphere as a thin layer show that the aerial perspective becomes an important one of many resources capable of pointing to this essentially invisible phenomenon.

The images gathered to point to climate change thus accumulate the history of using the aerial perspective in the environmental debate as negative images of past activities that have led to the problem. These are then put together with new images to indicate the current crisis. The focus here is on melting glaciers and collapsing ice shelves, images of hurricanes captured from space, of flooding (particularly in New Orleans, but also elsewhere), images of the destruction of the Amazon rainforest, and images of forest fires. Together they add up to an elemental combination of earth, water, wind, and fire.

As is common when a recognizable set of iconic images accumulates within a genre or subgenre, there is also a process of reconciliation perceivable in the ways in which the images balance socially antagonistic forces (Altman, 1999). The reconciliation is between the critics of postwar planning, who argue that what is needed is a dismantling of industry and the rejection of consumer capitalism, and those who argue that the issue is a matter of moving from old, dirty, polluting technologies—the carbon economy—to new "bright green" ones, or renewable energy sources. Thus the aerial image is joined in significant ways with the close-up, acknowledging the debates within the environmental movement about the need to change the relationship between human society and the earth. Finally, both aerial images and close-ups are put together with images that indicate the possibility that technology, after all, can be deployed both to communicate the problem to a wider public and to coordinate a solution. Here images of the earth from space play a critical role as emblematic of human achievement, of the possibility for contemplation, and of the slow cumulative process of data collection and modeling that has provided the scientific data indicating global warming. The contemporary icons of the computer and the earth from space regenerate the aerial perspective as key in the management and reversal of global warming.

The release of the documentary film biography of the U.S. politician Al Gore, *An Inconvenient Truth* (David & Guggenheim, 2006) has often been hailed as a turning point in the struggle to create a meaningful mass public discourse on climate change. The film provides a clear and accessible explanation of the history of empirical investigations that have led scientists to agree, first, that the earth's atmosphere is warming, and second, that human activities, particularly the burning of fossil fuels and the clearing of forests, are causing atmospheric warming beyond natural cycles. The message of the film is that the science is clear, a point underlined by the many eloquent explanatory images and the articulate contributions from researchers. The further point is that climate change will disrupt—is already disrupting—life on earth.

Although Gore argues in the film that the issue of climate change is moral rather than political, the message *is* clearly political, as Pat Aufderheide pointed out in her review of the film for *Cineaste* (Aufderheide, 2006). The political situation is unusual, however, in that *An Inconvenient Truth* constitutes a turn toward grassroots politics on the part of a career politician and can be seen as part of a much larger trend, using film to express oppositional messages during what Douglas Kellner defined as the Bush-Cheney era (Kellner, 2010).

Much has been written about the success of the film in the United States and globally, the most significant point here being its status in 2006 as the third most successful documentary in history. Several aspects of the film account for this kind of success on release in the cinemas and the continued significance of the film in the popular imagination through DVD sales, and through citation in further documentaries and other publications, as well as through spoofs and references in films such as *The Simpsons Movie* (Silverman, 2007). At least part of it must be attributable to the long development of the material over a number of years so that the film, adding its own layer of material and framing, is merely one stage of several stages of development that, as Gore puts it in the film, comes from an active interaction with scientists and members of the worldwide audiences of the slide show. About one hour and fifteen minutes into the film, Gore explains that he was "trying to identify all those things in people's minds that serve as obstacles to them understanding this, and whenever I feel that I have identified an obstacle I try to take it apart, roll it away, move it, demolish it, blow it up. I set myself a goal" (David & Guggenheim, 2006).

This process of clarification delivered through engaging rhetoric is accompanied by all kinds of different visual materials, documentary film and photography, graphs, animation, computer-generated images, and modeling, all enabled by the flexibility of the PowerPoint presentation format. These materials are further enhanced by the techniques developed by Davis Guggenheim to integrate the material into the film, one aspect being a particular emphasis on the quality of the images, making them as spectacular as possible. Another aspect involves the framing of the lecture as a process within the life of the lecturer, Al Gore, the ostensible subject of the film. This process allows the images to function not only as carriers of primary information but also as talismans, objects to hold on to and value for their own history and potential future. As part

The Visual Rhetoric of Climate Change Documentary 103

of Gore's store of accumulated virtual objects on the subject of climate change, the images take on a personal tactile quality and can thus be seen not as products of a detached process of scientific analysis but rather as integral parts of a personal journey.

In the process, communication about climate change is taken in a direction that includes different and potentially conflicting perspectives on climate change, reconciling or at least balancing the arguments for an embodied approach that recognizes the human being as natural within nature, with instrumental reasoning about solutions to physical problems.

In *An Inconvenient Truth* the reconciliatory movement between the close-up and the planetary perspective can be perceived at the beginning of the film, underlining the ways in which the two points of view are to be joined as part of one argument rather than pitted against one another. The very first image, accompanied by a simple two-note chord hit high on the piano, is a close-up of a leafy branch hanging over a slow-moving river, with the voice of Al Gore commenting. As the camera pans across the leaves, over the river—showing the sunlight bouncing off the brown water onto the tree trunks, the leaves reflected in the water broken by half-submerged plants and rising mist—the accompanying words emphasize the integration of the senses of sight, hearing, and touch:

> You look at that river, gently flowing by; you notice the leaves rustling with the wind, you hear the birds, you hear the tree frogs. In the distance you hear a cow; you feel the grass; the mud gives a little bit on the river bank. It's quiet, it's peaceful, and all of a sudden it's a gear shift inside you, and it's like taking a deep breath and going [a deep breath can be heard], "Oh yeah, I forgot about this."
>
> (David & Guggenheim, 2006)

Immediately following this sequence, about a minute and a half into the film, viewers see an image of the earth within a frame. It is a close-up shot of a laptop computer showing the image of the planet earth known as *Earthrise*. The camera zooms into the frame until it takes up the whole screen, and the film itself appears to be showing viewers planet Earth as it looks from the moon—and the voice is now not contemplative but engaged in the act of public speaking. The camera pans over the faces of a listening audience in close-up while Gore explains what the image is and where it comes from.

The movement in and out of the image of the earth, its containment within the frame, the placement of Gore in front of the image lecturing on it, and the concentrated faces of the audience members all bring the image into the sphere of communication rather than into the quasi-religious contemplative sphere of reverence and wonder, which has now passed to the trees, plants, and animals. Although it never quite loses its spectacular perspective, the whole earth image becomes an accessible, touchable image that can be manipulated and adapted as well as wondered at, accessibility that Gore then demonstrates through a playful series of scientific visual achievements through remote sensing, such as an earth without clouds created out of 3,000 digitized images, and a spinning earth made through time-lapse photography that shows one day of rotation compressed into 24 seconds.

This spinning earth is seen again in the final chapter of the film, in which Gore explains the "pale blue dot," an image of the earth taken from a space probe 4 billion miles out in space. As the camera zooms in on the projected image, Gore's comment combines a statement about a digital image, a pixel, with a statement about historical perspective, all human history. "Everything that has ever happened in all human history has happened on that pixel" (David & Guggenheim, 2006). The image changes from this potentially unsettling perspective, in which the significance of individual existence threatens to completely disappear, to the Earthrise image of the opening, followed by the "blue marble," the laptop and the speech venues, the flight over the ice floes and the mountains with the helicopter blades faintly audible, and then finally to the slow-moving river, the camera now panning back from right to left toward the river bank and the sun. The symmetry of the opening and closing maintains the balance between the planetary and the personal perspective, between the world as image and as home, experienced through all the senses.

Within this frame of the personal and the planetary, the film contains the lecture, itself an example of classical rhetoric. The lecture is testimony to the continued strength of the humanist tradition of rhetoric in North American political discourse. The five elements of rhetoric, *inventio* (defining the subject), *dispositio* (laying out the arguments), *elocutio* (choosing the words), *memoria* (memorizing the speech), and *actio* (delivery of the speech), are all portrayed as having been mastered over a long and arduous process of self-education (see Minster, 2010). The structure of the lecture also follows the rhetorical model: the initial demonstration of personal integrity and commitment ("I used to be the next president of the United States"; archive television images of Gore campaigning during the 2000 presidential elections, often referred to as "the stolen election"; images of Gore continuing his political life through his campaign on global warming); the appeal to the sense of community (images of audiences from different parts of the world; the history of the "whole earth" images produced as part of the space program); the presentation of evidence and the integrity of the sources (references to teachers and friends who are scientists); the refutation of detractors (demonstration that the oil industry deliberately framed global warming as theory rather than fact); and the final intensification and call to positive action (list of examples of collective success from defeating Nazism in World War II to closing the hole in the ozone layer in the 1990s; "We have the ability to do this. Each one of us is a cause of global warming but each of us can make choices to change that.")

The images that accompany the lecture in the film are both evidence in themselves and in part what Keith Beattie captured with his analysis of the role of display in documentary film, recognizing that the appeal of documentary film lies as much in the pleasure of looking as in the desire to absorb the exposition: "Separated, or freed, from the immediate demands of knowledge production, documentary display entertains, startles and excites in ways which produce pleasure—the great repressed in analyses predicated on documentary as a sober discourse" (Beattie, 2008, p. 5). At different stages in the film, images are introduced in montages of the most visible consequences of global warming that are then repeated individually later in a more extended analysis of causes and consequences. Thus a 15-second sequence—consisting of flight over ice floes, shifting to sunlight from the bottom of a tall forest tree, then close-ups of melting snow followed by cracked earth, the top of an industrial building against the sky belching out smoke, and a field on fire—is inserted into a monologue in which Gore reflects on his failure to convey the message. These sequences are not about evidence, for there is no time or supporting information to give them that status. Nor are they about establishing a connection between people and the environment. They are pure spectacle, a dramatization of the message and the moment, an advertisement for the campaign. The trailer of the film comprises such an intense sequence of high-impact images that it comes across as hysterical. It is a relief that the whole of the film is not like this.

The film and the lecture make use of the spectacle of images in ways that reflect the rise of popular aerial photography in books like Yann Arthus-Bertrand's *Earth From the Air*, one of the most successful books to be published in 1999, a collection of aerial photographs taken all over the world (excluding Yemen, Saudi Arabia, and China) over a period of six years (1994–1999) and presented with texts relating to environmentalism, ecology, and sustainable development. Cosgrove and Fox argued that such books have made the connection between aerial images and environmental issues so close that they have gained dominance:

> Since the 1970s the aerial perspective seems to have become closely bound to environmental concerns, so that not only is the technology of aerial photography and remote sensing a principal

The Visual Rhetoric of Climate Change Documentary 107

> source of data for environmental monitoring and modeling of such phenomena as climate change, ice melting, vegetation loss and species extinction, but, in the popular imagination, aerial images of the earth's surface and landscapes have come to be framed almost exclusively through an environmental lens.
> (Cosgrove & Fox, 2010, p. 87)

Thus in many ways the film, like contemporary art photography, has to struggle against the unthinking nature of the spectacle that grabs attention but fails to move. In the discussion of the impact of fossil fuel mining, a key factor in global warming, photographs by Edward Burtynsky of opencast mining are used, themselves part of a contemporary photographic trend that has developed since the 1990s using large-format photography to portray in a stark "deadpan" and detailed way the realities that sustain modern urbanized mass culture (Cotton, 2009). Such images are carefully framed and startling in the quantity of visual detail they hold, but they do not demonstrate a struggle to make an impression, and the meaning of the images is left open. The blank, still quality of the photograph simply freezes its point in time and allows inevitable cognitive effects on the viewer.

Contrasting with this detached point of view, the most palpable consequence of global warming for American audiences is expressed in dramatic and highly emotionally charged footage of Hurricane Katrina as it hit New Orleans in August 2005 at the time *An Inconvenient Truth* was being made. Television reports of the disaster can be seen on the laptop screen in the film, the dampened voices of reporters and interviewees distantly recalling the horror of seeing the people of the city struggling to survive. The section of the film that treats the hurricane begins, however, with a sequence of satellite images showing three quarters of the earth and the hurricane in the form of a cloud perfectly swept up into a white swirl. After a sequence of images using infrared to perceive movement below, aerial shots move over the flooded streets of the city, observing people stranded on the tops of buildings. A voice speaking over a radio says, "I can't take it anymore. The water is up to my neck.

I don't think I can hold out." This sequence captures the strange and tragic concurrence of apparent total observation and understanding of the natural disaster through technology together with the impossibility of prevention or even adequate intervention. The argument is, of course, that prevention requires a long-term strategy to reduce global warming and thus the strength of the hurricanes, and that intervention requires careful acknowledgment of the dangers faced, planning, and the implementation of safety measures regardless of economic cost. It is a prosaic message in political terms, but in the context of Hurricane Katrina it takes on a kind of fervor that such messages seldom achieve.

One hour and 19 minutes into the film, Gore moves into the part of his lecture that involves the call to action. After showing the statement "Humanity already possesses the fundamental scientific, technical, and industrial know-how to address the carbon and climate problem" (Stephen Pacala & Robert Socolow, *Science*, 2004; quoted in David & Guggenheim, 2006) the image moves from graphs of incrementally achieved reductions in carbon emissions to black-and-white aerial footage of rows of housing, recalling images taken by William Garnett of housing being built in Los Angeles after the Second World War as part of Gutkind's project for *Our World from the Air* (1952) and as part of an environmental critique of such housing in the Sierra Club publication, *This is the American Earth* (1960). These images of mass housing followed by aerial images of densely packed streets and slow-moving car traffic are counterpoised by the words of the lecture: "Each one of us is a cause of global warming, but each of us can make choices to change that" (David & Guggenheim, 2006). In this way the history of such images is shifted so that they are neither about the manufacture of mass happiness nor about people being planned by higher forces into a spirit-stifling mass grid. Instead, the images become part of a longer history of growth and individual prosperity that has reached a point at which choices can be made "with the things we buy, the electricity we use, the cars we drive. We can make choices to take our individual carbon emissions to zero" (David & Guggenheim, 2006). The images of mass housing and motorized transport change to images of new grids and crowds—solar panels, an electric railway, a new low-emissions car, people on bicycles, and a row of wind turbines along a green ridge.

This vision of the solution to global warming has been heavily criticized. Timothy Luke, for example, argued that it does not present an argument for any real change from the current model of "natural capitalism" but rather provides "anchors for a new revitalized regime of sustainable degradation forged in response to the current conditions of capitalist production" (Luke, 2008, p. 1818). This argument would support the idea in the debate about solutions that there is no real reconciliation between the perspective of deep ecologists and the more instrumental approach of the "bright green" arguments, and that the montage of aerial images described earlier still marks a point of divergence rather than convergence.

Luke also argued that Gore's claims about the failure to acknowledge climate change are a willful distortion of history in order to "build a sense of drama about global warming":

> Gore's own exile in the wilderness outside of Washington's 24/7 news cycle ... is used to exaggerate the degree to which global warming has been ignored in the USA. When coupled with the unwillingness of either President Bush [George H. W. Bush, 41st U.S. president] or Bush [George W. Bush, 43rd U.S. president] to enter into a serious debate about climate change, and a propensity at the White House to harass scientists in government service who study global warming, there is an illusion of inattention to climate change that Gore depicts as smug complacency unable to face "an inconvenient truth."
>
> (Luke, 2008, p. 1815)

This argument is about the awareness of the general public and, perhaps more significantly, the willingness of business corporations to acknowledge the issues. As Aufderheide put it in her critique of *An Inconvenient Truth*:

> The crippling of public debate by the mainstream news and the crippling of political action by lobbyists are not only worth knowing, they are crucial pieces of the explanation for how an ostensibly capable public—a public in which Gore now places his trust—could be so deluded and delayed in its response to a crisis barreling down upon us."
>
> (Aufderheide, 2006, p. 55)

The aerial view is thus integrated in *An Inconvenient Truth* into the personal journal of a politician and climate change activist associated with lobbying and campaigning for the transition to renewable energy sources. Its integration of the personal perspective into the political overview is part of a conscious strategy to reach out to a grassroots perspective, but it is clear that the starting point for that is professional politics. In contrast, *Everything's Cool* represents a more genuine example of grassroots documentary filmmaking.

Although it went into production before *An Inconvenient Truth*, the documentary *Everything's Cool*, directed by Daniel B. Gold and Judith Helfand, appeared in 2007, the year following the other film's release. Like their previous film, *Blue Vinyl* (Gold & Helfand, 2002), this documentary is marketed as a "toxic comedy" that asks in a jaunty, open, and engaging way why the general public in the United States has failed to engage the issue of global warming. On the cover of the DVD and on its posters is a comic montage of iconic images associated with climate change campaigning and disaster movies with a global warming theme —two penguins stranded on a piece of floating ice, the White House half submerged in the ocean, a heavy sky with forked lightening cracking down on the presidential residence. Along the way, this film asks explicitly whether there is something wrong with the environmental movement, bringing together frustrated and even depressed activists who have been working on the issue for decades (Ross Gelbspan, Bill McKibben, Heidi Cullen, Rick Piltz), with slick newcomers (Michael Schellenberg, Ted Nordhaus) who argue for a fresh new direction. Thus it portrays policy formation in action through dialogue between activists of different generations and between activists and the general public. Consequently, it has very little use for aerial images but does use them in a significant way at a single point in the film.

The film begins with a spoken statement reinforced first by a pixilated sequence in which boxes of papers accumulate on a bench until the frame has been filled and then by the appearance of some of the words on-screen synchronized with the voice (underlined words in the following quotation are spoken and also appear on screen; the other words are only spoken):

> At the turn of the twenty-first century, several thousand tons of scientific studies on climate change all led to a single revolutionary conclusion. By burning fossil fuels human beings are changing the climate (a lot). And if we keep doing that the planet could become pretty much uninhabitable.
> (Gold & Helfand, 2007)

The film continues with a statement of its mission:

> In the fall of 2004, at a moment when polls showed that global warming ranked dead last as a voting issue, we rented a biodiesel-ready but usually gas-guzzling truck and plotted out a cross country tour to see if the polls were right.
> (Gold & Helfand, 2007)

Begun before *An Inconvenient Truth* and including footage of Al Gore's lecture before the film version was made, *Everything's Cool*—described by reviewers as "lighthearted" (Foundas, 2007), "witty" (K. A. MacDonald, 2007), and "breezy" (Holden, 2007)—uses the jaunty and engaging tone of activist documentary, much influenced by the work of pioneers like George Stoney and Michael Moore, and typically creating a community event around which to conceptualize its message—the journey around America in a lorry showing a huge blown-up image of the cracked earth already mentioned. This low-budget indie film, again, uses only 13 seconds of aerial footage—namely, a shot of burning rainforest, two shots of New Orleans after Katrina, a shot of a plant, and an establishing shot off the coast of Alaska.

Most of the footage of the film is thus of people in medium shot or close-up talking, in interviews or in groups, or of people carrying out tasks such as making fake snow, concocting biofuel, or taking part in community action. This reflects the overall approach of the film, which is to show the difficulties of media communication, compared with the ease of person-to-person contact. Thus the film tells the shocking stories of media manipulation and of the gagging of scientists, capturing the struggles of individuals such as Dr. Heidi Cullen of the Weather Channel in her efforts to learn how to combine science with the strange skills of media presentation. The film also shows that it takes a major natural disaster—Hurricane Katrina—for the media to play its role in "closing the gap" between scientific knowledge and public awareness, confirming the points made by several media analysts that the events-led nature of public media poses a problem for climate change communication policy (Doyle, 2007; Kitzinger & Reilly, 1997). The principal message of the film, then, is to point to community action and person-to-person communication as the best strategies for communications policy on climate change. What is more, the film seems to say that this in itself is already part of the solution to the climate crisis.

The one sequence in which traditional establishing aerial shots were made specifically for the film is worth examining in detail for the way it contrasts with the strategy used by the other two films discussed here. It begins with an archive shot of some refreshingly dirty ice cliffs falling into the ocean. A violent storm driving waves onto the land, aerial shots of burning forest, shoulder-height shots of emaciated cattle and close-ups of dried-out earth quickly summarize the point that "by 2005 the physical evidence of global warming was physically outpacing even the most alarming scientific predictions" (Gold & Helfand, 2007). Aerial shots of U.S. industry then cut to shots from a light aircraft used to access the location in Shishmaref, Alaska, that the team is to visit, an Inupiate village 80 miles east of Siberia, "a place where you could already see global warming happening" (Gold & Helfand, 2007). This shot is integrated into the personal gaze of the filmmakers through the commentary that states, "We wanted to see the impact of America's inaction" (Gold & Helfand, 2007). Seeing global warming happening is translated visually into a shot of a small wooden building hanging over the edge of a frozen cliff battered at its base by relentless waves. These images are clearly not an attempt to dazzle the viewer into paying attention because they cannot be classified as sublime or beautiful even though they are dramatic. Dean Kuzuguk, a Shishmaref resident, indicates what permafrost is, pointing to a low bank of mud that forms the coastline. Kuzuguk then shows where the cliff-top houses used to be, naming their owners before Leona Goodhope takes over the story by pointing out that the place where her own house used to be is "completely gone." She repeats the phrase, apparently finding it difficult to comprehend the fact herself, even though she can see it. Judith Helfand interviews a group of children on what they would take with them if their whole village was moved, as has happened in this case, and their answers—some clothes, the dog, something to eat, a Game Boy—demonstrate the ways in which children comprehend such events. The emblematic image that counts as physical evidence of climate change here is not a dramatic and colorful aerial shot but an extreme close-up on a drop of muddy water accumulating on the tip of a small brown root that pokes out of the coastal overhang.

The last third of the film deals with attempts to renew the energy for environmental campaigning in the face of the opposition, inertia, and resignation to the physical evidence that has been diagnosed. These are couched in terms of potential opposition between the established environmental movement and newer strategies—sometimes grouped under the label "bright green." Bill McKibben, author of *The End of Nature*, gives an account of his attempts to get people to hear his message and in the process introduces his vision of enjoying the present:

> In the course of doing the work that needs to be done, don't divorce it from the living that needs to be done too. Right now is the time to be out there in it, in the human communities, in the natural communities that are still in their great glory. All of which is a long way of saying that I'm going out skiing this afternoon [laughter].
>
> (Gold & Helfand, 2007)

Public relations consultant Michael Schellenberg and political pollster Ted Nordhaus, self-published authors of *The Death of Environmentalism: Global Warming Politics in a Post-Environmental World*, are introduced as oppositional figures who clearly reject the established visual iconography of climate change campaigning: To the film audience Schellenberger says, "I would suggest that it would take more than dead penguins and melting ice caps to get Americans to fundamentally get involved in this political transformation of our energy economy" (Gold & Helfand, 2007). Ziser and Sze have analyzed the portrayal of Schellenberg and Nordhaus's "breakthrough" agenda in the film as a negative one, in which their "glib technophilia" is revealed as distant from "the complex, historically determined reality of the people on whose behalf environmental decisions are being made" (Ziser & Sze, 2007, p. 403). Their vision, however, is quickly integrated in the film into the process of renewal, so that their work with communities becomes part of the new vision that includes both the old and the new environmentalists in a program for new jobs:

> The great news is that when we make the transition to a new energy economy we are going to create three million jobs in the United States, we are going to free ourselves from foreign oil and we are going to clean up the air, and it's a very positive vision.
> (Schellenberger; quoted in Gold & Helfand, 2007)

The tragedy of a quarter century of failed efforts at media communication is captured at a rally at the end of the film. The event constitutes a turn to a traditional form of political communication in Vermont, a five-day march on the capital city: "Their strategy: Use some of the momentum from mass citizen action to challenge politicians to support the most far-reaching emissions reductions ever proposed in the United States" (Gold & Helfand, 2007). In a speech to the gathered walkers, Bill McKibben speaks of the therapeutic nature of community action that is needed to sustain both the science and the politics. "The one part of the environmental movement that has been missing is the movement," he states, "to think and work about global warming can be a difficult business. This was the most hopeful day for me of the last twenty years" (Gold & Helfand, 2007). His argument that "the economy feels more real to people than the physical world" is dealt with through the active expression of the bodies of the walkers and demonstrators.

In contrast, *The 11th Hour* (Conners Petersen & Conners, 2007), produced and presented by Hollywood actor Leonardo DiCaprio and codirected and cowritten by Leila Conners Petersen and Nadia Conners of Tree Media Group, is generally seen as aimed at a young audience heavily invested in media and consumer society. The pace of the film is much faster than that of either *An Inconvenient Truth* or *Everything's Cool*. A true product of the information society, the film uses an astonishing amount of archive imagery and a bewildering number of interviews with different thinkers on ecological and environmental issues. Nevertheless, the excess forms into a stream of collective thought that coalesces around agreed theses on the physical causes and consequences of global warming, the psychological causes, the economic issues, and the design solutions. The interviewees are placed to the right of the wide screen so that their names and qualifications can be clearly indicated to the left. Their summaries and explanations are illustrated by footage directly related to their themes. Occasionally the speech pauses and the images permit a degree of contemplation. The film manages almost miraculously to create an intelligent and personally engaging coherence from the sheer energy of the different individuals the filmmakers have involved and the various highly developed personalities, perspectives, theories, and activities that are combined.

Not surprisingly, the role of aerial footage in this remarkable effort at cohesion is also varied and interspersed with all kinds of spectacular imagery of the earth and of human activities. The beginning of the film is atypical of the whole; an accelerating montage of images somewhat in the style of *Koyaanisqatsi* (Reggio, 1983), the sequence is a reflection on the relationship between the earth and human history. Two seconds of footage of a tornado over a field is followed by darkness and then two seconds of a fetus in the womb; after more darkness comes a sudden flash of a storm, then a turtle flying underwater, a flash of the storm on the gulf coast, a tracking shot along an ice wall, an aerial shot flying just above a mountain forest, a car overturned by winds, the fetus in the womb, people running down a street; and then the sun, eclipsed by the moon, setting in time lapse; a passage down a wind pipe to the sound of breathing; a fireball, rising smoke, oil extraction, a quarry, an aerial shot of a complex of refineries, a man being searched, an exhaust pipe of a car, a car overturned, a starving man, floodwater running down a street, a car falling over a cliff, a chimney being detonated, a city in a storm, black-and-white footage of a nuclear explosion, a missile being launched, military maneuvers, grazing animals gathering, a piece of meat being cut, fork lightening, a bear roaring, smoke from a wide chimney, a row of chicken carcasses moving along in a factory, people moving down the aisle of a factory, lava flow, a building exploding, a polar bear moving among some rubbish—and so on and so on, the images coming closer to the themes of global warming as the montage continues.

This montage speaks of the endless variety of striking sights that have been framed and captured of the earth and the experiences of human beings. The range of possibilities for image making demonstrates that no particular perspective is intrinsically more spectacular than any other; all are spectacular, no part of the earth is inaccessible, and the visibility of the earth is from within and from without, from above and below, along with and up against. The dazzling variety of voices from different communities expressing different ideas but all somehow existing as part of a whole argument parallels this cornucopia of images.

The Visual Rhetoric of Climate Change Documentary 117

Released after *An Inconvenient Truth*, the later film encountered flak from the popular press relating to celebrity lifestyles and green fatigue, but as Tom Meek pointed out, the film makes two clear and serious points: first, that "global warming is more a socio-political issue than a scientific conundrum" (Meek, 2007, p. 58), and second, that the key to survival lies in developing techniques to live on the energy provided by the sun in the present rather than using up an ancient store and releasing carbon into the atmosphere. This brings up a number of intriguing new approaches to energy production and materials science that the extras on the DVD follow up. Like *An Inconvenient Truth* and *Everything's Cool*, *The 11th Hour* employs cohesion, creativity, and above all unity in diversity; this is the strong strategy of this visually energetic film.

Conclusion

There is no lack of aerial images to assist in the communication of climate change policy and by now (2012) it has become normal for each climate change–related event to be reported using aerial shots. The use of these in documentary films is, as has been shown, part of an overall strategy aimed at drawing in and engaging audiences. The argument that the aerial perspective needs to be embedded in a more person-centered strategy is one that comes across strongly in all three films, notwithstanding the extent to which each of them uses, ironizes, enhances, or rejects the spectacular quality of images that are available. Thus although a single chapter does not afford sufficient space to fully analyze every use of this perspective in the three films, I argue that the use of the aerial perspective in each of these contemporary climate change documentaries has been part of reaching out to a particular audience in a way tailored to meet its needs, representing carefully collected and manipulated data (*An Inconvenient Truth*), exploiting the emotional response to the sublime (*The 11th Hour*), and establishing the localized impact of climate change on a particular community (*Everything's Cool*).

The representation of collected data is an aspect of aerial photography that is being extended through the use of computer-generated imagery (CGI) to project into the future. The ironies generated by the use of aerial photography and digital technology in warfare and surveillance are evident. Analyzing Farocki's *Images of the World*, for example, Allan James Thomas wrote that the film "returns repeatedly to the role of the image in transforming phenomena into data; the analysis of the movement of water in an artificial wave tank, photographic scale measurement, image processing, military aerial reconnaissance, police identikit portraits, architectural modeling" (Thomas, 2002). For the climate change documentary, too, the aerial image is data. Earl B. Brown described the revolution in optical engineering back in 1967:

> The key to current methods of analysis is the application of communication theory to optical systems. Around the early 1950's it began to be recognized that an optical system is a communication system, and that it may be analyzed in the same manner in which electronic systems are analyzed. Fundamental to this is the concept of a photographic scene as a "message" composed of a collection of signals over a range of frequencies—in this case, *spatial* frequencies.
>
> (Brown, 1967, p. 103)

The representation of the accumulation of images of the world as the creation of a vast information database over time also confirms Cosgrove and Fox's definition of the purpose of photography as "at once prosthetic and aesthetic" (2010, p. 8). For communicators of climate change issues this is a critical point. Documentary photography is attractive and compelling in itself, and at the same time it represents the capacity for technology to extend the senses and knowledge over space and time. More important, Cosgrove and Fox moved from this point to express a further significant aspect of the technology of the photographic that is an integral part of the development of image consumption as part of communications policy. They elaborated on the purpose as not only to increase the sphere of what can be seen but also to enhance what can be

The Visual Rhetoric of Climate Change Documentary 119

done with the image once achieved: "to capture and freeze a moment in space and time, documenting and archiving it, and rendering it mobile through the printed or transmitted image" (Cosgrove & Fox, 2010, p. 8).

Most of the photography and film footage used in the documentary films discussed here, particularly of events such as Hurricane Katrina and the collapse of ice shelves, is archive material; the same material is sometimes used in more than one of the films. This "recycling" demonstrates the ways in which, once recorded, images can contribute to many different projects in a variety of ways to help reflection on their meaning in the broader social context.

The extension of the capacity to see the planet from ever greater distances through technology produces in viewers "powerful phenomenological effects" (Cosgrove & Fox, 2010, p. 9). Environmental documentary film producers have recognized the potential of these effects to capture the attention of viewers, to win them over to documentary film, and to persuade them not only of the urgency of the situation regarding climate change but also of the possibilities for responding individually and collectively to it.

This development has been underpinned by developments in environmental activism that have rejected the binary, divisive tendencies of ideology and widened the public perception of expertise to include not only scientists, managers, and politicians but also farmers, agricultural workers, activists, filmmakers, photographers, and indeed the viewers themselves, who take on some of the discourse of management in combination with their own specialized, localized view, to represent an environmentally informed perspective (see Beck, 2007; Heller, 2005). The emphasis on community involvement in filmmaking, particularly in Gold and Helfand's *Everything's Cool* and in Connors Petersen and Conners's *The 11th Hour*, links the range of images (from close-ups to aerial shots) with the continuum of possible human and technologically extended perspectives to represent not just an emotional image—beautiful or sublime—but an embodied statement about calculable value for communities in terms of productivity, sustainability, and well-being.

The significance of using aerial footage in environmental campaigning material is the very fact that it is admitted and becomes fully integrated into the point of view of a community-inspired environmental documentary vision. The aerial view remains a signifier of grandeur and of "one world," and it also retains its role as surveillance footage—but now, in addition to its military purposes and its use in spying activities, it is also a means to reveal the impact of human activity on the environment and to show the finitude and vulnerability of the planet. These impacts and the now iconic aerial imagery used to signify them can be listed as consciously chosen elements of a climate change documentary subgenre: images of the earth from space, of melting and collapsing ice, of burning rainforest; images of industrial complexes involved in the extraction and consumption of fossil fuels; images of urban sprawl, of hurricanes and flooding; images of the earth being observed and analyzed; and finally, the new objects on the horizon, images of alternative energy production and images of mass demonstrations and community walks in favor of political change.

Part II

Public Engagement Through Public Participation

CHAPTER 5

STATE COMMITMENT TO PROMOTING PUBLIC PARTICIPATION

THE UNFCCC AND CITIZENS' ROLES IN CLIMATE CHANGE POLITICS

Anabela Carvalho
and Joyeeta Gupta

INTRODUCTION

Given the nature of climate change, there is no doubt that citizens need to engage with the issue at many different levels. On the one hand, changes in individual behavior are necessary in order to address the problem. On the other hand, citizen views are a fundamental basis for any policies and measures adopted by a government and can lead to both political action and inaction. Most often such views are elicited through surveys and similar instruments for gathering and aggregating data on "public opinion." However, these proxies for the public sensibility on climate change are far from sufficient for developing effective responses to climate change. Citizen engagement with the policy-making process is vital. In order to gain legitimacy and public acceptance, policy decisions need to be made through a process that is perceived as fair. That requires inclusive participation of different sectors of society as well as thorough accountability. Moreover, citizens arguably are entitled to contribute to identifying and choosing options to address climate change, and it has been shown that although it is challenging, public participation can improve the acceptance and (according to some) even the quality of political decisions.

Wider political engagement with climate change through political action outside of or beyond the spaces created by governments is necessary in order to address climate change, but in this chapter we focus on public participation promoted, initiated, or invited by states: "organized processes adopted by elected officials, government agencies, or other public- ... sector organizations to engage the public in environmental assessment, planning, decision making, management, monitoring, and evaluation" (Dietz & Stern, 2008, p. 1). Under the United Nations Framework Convention on Climate Change (UNFCCC), signatories have committed to promoting public participation in climate change politics and to regularly reporting on their achievements. What have states done in order to engage citizens with climate change and promote public participation in policy processes? How do existing legal frameworks define states' roles and responsibilities in relation to public participation in climate politics, and how have countries been implementing such responsibilities? How are citizens and state–society relations constituted through the discursive work of state reporting to the UNFCCC? In this chapter, we address these questions through the analysis of the national communications (NC) reports of six countries with different levels of greenhouse gas emissions and vulnerability to climate change.

State Commitment to Promoting Public Participation 125

The chapter starts with a discussion of research on public participation. It then moves to a description of the legal and political framework for states' commitments on public engagement with climate change as set up by the UNFCCC and additional documents. The following sections offer details of the design and research methods employed in this study and discuss the findings from the analysis of NC reports to the UNFCCC by six countries.

RESEARCH PERSPECTIVES ON PUBLIC PARTICIPATION

In the last few decades, public participation in policy processes has become common currency in the discourses of official bodies, nongovernmental organizations (NGOs), and academics. Dwindling public confidence in politics, the appearance of new complex or "wicked" problems, and the development of international agreements recommending (or determining) the need to hear the public have led multiple voices to call for involving citizens in decision-making processes in areas such as land-use planning, transportation, and the environment.

Various sets of ideas in academia have propelled the popularity of publication participation. Works on democratic theory and philosophy produced since the 1960s have favored participatory approaches to decision making (e.g., Barry, 1965; Rawls, 1971). The theory of deliberative politics put forth by Habermas (1996) and others suggests that public argumentation and deliberation can produce rational consensuses and lead to more qualified decisions. This is often thought of as an ideal situation in which citizens would have equal access to political processes and in which they would be able to participate without constraints. The possibility of unhindered expression and the universal commitment to truth would also be required. Even though it is unrealizable, several authors have defended the use of this ideal as a yardstick to evaluate "deliberative politics in action" (Steiner, Bächtiger, Spörndli, & Steenbergen, 2005). Dryzek's works on discursive (1994) and deliberative (2000) democracy have also contributed to placing citizenship and public discourse on the agenda. With more direct relevance for problems like climate change, Beck's theory of "reflexive modernization" (Beck, Giddens, & Lash, 1996) values public participation in policy processes in today's risk society as traditional political institutions become irrelevant or inadequate to dealing with new challenges.

Academic literature on policy studies offers various understandings of public participation. The degree of openness of policy processes, the stages at which the public is or should be allowed in, the procedures for selecting the participating public, the methods of public involvement, and the implications of the outcomes of participation for decision making are all matters of normative contention. In her seminal work, Arnstein (1969) argued that true participation involves a high level of empowerment of the public with input into the decision-making process, and she called processes such as information and consultation tokenism. Empirical research on public participation has focused on a wide variety of designs and mechanisms of participation, and it is difficult to draw strong conclusions from existing studies.

Viewed as "the practice of consulting and involving members of the public in the agenda-setting, decision-making, and policy-forming activities of organizations or institutions responsible for policy development" (Rowe & Frewer, 2004, p. 512), public participation is considered by many desirable for a number of reasons. Normative, substantive, and instrumental rationales (e.g., Stirling, 2008) for citizens' participation in the governance of public problems have been presented. Public participation is said to increase the legitimacy of decision making because issues are more widely debated and at least some degree of public agreement is likely to be reached (Fischer, 2000). Such legitimacy requires that processes be perceived as fair in terms of access and power to influence the outcomes (Webler, Tuler, & Krueger, 2001). Public participation is also lauded for increasing the accountability of the decision-making process as more (and more diverse) participants experience it from the inside and have a say in at least some of its phases. Some have maintained that the inclusion of diverse viewpoints, arguments, and forms of knowledge can widen the definition of problems and improve the quality of the decisions, a substantive justification for participation of the public. In other words, public participation can arguably help avoid "formulating the wrong problem by incorrectly accepting the false metahypothesis that there is no difference between the boundaries of a problem, as defined by the analyst, and the actual boundaries of the problem" (Ulysses, 2012, n.p.). Increased public acceptance and trust may also result from participatory exercises. This may help avoid conflict in the implementation phase because the public is more likely to accept decisions resulting from a process that is viewed as democratic and inclusive. More generally, public participation can raise awareness of collective problems and offer opportunities for learning (Huitema, Cornelisse, & Ottow, 2010).

In the area of science and technology studies, the notion of public engagement with science and technology (PEST) has become quite influential in the last two decades or so (e.g., Felt & Fochler, 2008; "From PUS to PEST," 2002). In contrast with previous understandings of citizens as passive recipients of scientific knowledge and a focus on dissemination of information (the so-called deficit model), the current emphasis is on engaging the public at multiple levels in the governance of science and technology and on the promotion of more deliberative forms of decision making. Nonexpert forms of knowledge are now viewed as valid, and the dialogic interaction between researchers and citizens is praised, even though so far it has rarely fulfilled its promises (Phillips, 2011).

Despite these merits, participatory processes have been criticized for a series of shortcomings that impede the realization of their democratic potential:

> The limitations, deficits and constraints of participatory exercises, specifically as they are built into the respective format of the exercise, may concern the range of participants, the scope of the

agenda, the formulation of policy problems, or the stage of the policy process at which participation comes into play.
(Braun & Schultz, 2010, p. 406)

Assumptions regarding people's willingness and aptitudes to participate may be unfounded. Their opinions may be easily manipulated or they may tend to avoid conflict and confrontation (Van de Kerkhof, 2006). As Van de Kerkhof (2006) maintained in a study on climate change options in the Netherlands, policy makers are often reluctant to integrate the recommendations or preferences emerging from participatory processes in policy plans and perceive those exercises as "a communication tool, to educate the stakeholders ... rather than a process that produces useful insights for policy making" (p. 297). Finally, participatory processes have been considered very costly in terms of the required time and resources (Powell & Colin, 2009).

Although these problems must be taken seriously, they should not lead one to abandon public participation, especially because systematic research on the effectiveness of participation exercises is lacking. As noted earlier, most studies are not comparable owing to variations in design, topics, and methods. A structured research program needs to be developed and implemented (cf. Rowe & Frewer, 2004).

In what has been called a "participatory paradox" (Powell & Colin, 2009), public participation is in most cases initiated by an official body, governmental agency, or research institution. Such is the nature of the issues under examination in this chapter. There has been much debate on the pros and cons of top-down sponsoring of participation, and it is not clear how meaningful and effective these forms of engagement are. The consensus orientation of formal participatory exercises may annihilate more marginal views and produce false agreements. Ascertaining these matters is not the goal of this project. Instead, we analyze how the relationship between the state and the public is constructed in discourses on public participation. As Braun and Schultz (2010, p. 406) argued, "'the public' ... is never immediately given but [is] inevitably

State Commitment to Promoting Public Participation

the outcome of processes of naming and framing, staging, selection and priority setting, attribution, interpellation, categorisation and classification." Here we look at legal and policy discourses on public participation and examine the status and identities that are assigned to citizens and to the state. In the next section, we focus on the legal and political dispositions regarding the promotion of public participation in climate politics and start identifying processes of social positioning.

INTERNATIONAL COMMITMENTS TO PUBLIC PARTICIPATION IN CLIMATE POLITICS: THE LEGAL AND POLITICAL CONTEXTS

In international relations theory, regimes are sets of "principles, norms, rules, and decision-making procedures around which actor expectations converge in a given issue-area" (Krasner, 1982, p. 185). In the climate regime the main reference regarding matters of public participation is the UNFCCC, agreed to in 1992, and particularly its article 6, although there are further significant developments.

Signatories of the UNFCCC have committed to objectives related to informing, educating, and involving the public in relation to climate change. Such commitments are introduced in article 4 of the convention.

> Article 4
> COMMITMENTS
> 1. All Parties, taking into account their common but differentiated responsibilities and their specific national and regional development priorities, objectives and circumstances, shall:
> (i) Promote and cooperate in education, training and public awareness related to climate change and encourage the widest participation in this process, including that of non-governmental organizations.
> (UNFCCC, 1992)

Although the commitments it sets up for states include the promotion of public access to information and public participation in developing responses to climate change, article 6's heading privileges the domains of education, training, and public awareness. It defines the goals of the parties to the convention as follows:

> Article 6
> EDUCATION, TRAINING AND PUBLIC AWARENESS
> In carrying out their commitments under Article 4, paragraph 1 (i), the Parties shall:
> (a) Promote and facilitate at the national and, as appropriate, subregional and regional levels, and in accordance with national laws and regulations, and within their respective capacities:
> (i) the development and implementation of educational and public awareness programmes on climate change and its effects;
> (ii) public access to information on climate change and its effects;
> (iii) public participation in addressing climate change and its effects and developing adequate responses; and
> (iv) training of scientific, technical and managerial personnel;
> (b) Cooperate in and promote, at the international level, and, where appropriate, using existing bodies:
> (i) the development and exchange of educational and public awareness material on climate change and its effects; and
> (ii) the development and implementation of education and training programmes, including the strengthening of national institutions and the exchange or secondment of personnel to train experts in this field, in particular for developing countries.
> (UNFCCC, 1992)

The Kyoto Protocol to the UNFCCC, agreed upon in 1997, reiterates commitments on education, training, public awareness, and public access to information but leaves public participation out.

> Article 10
> [All parties shall:]
> (e) Cooperate in and promote at the international level, and, where appropriate, using existing bodies, the development and implementation of education and training programmes, including the strengthening of national capacity building, in particular human and institutional capacities and the exchange or second-

State Commitment to Promoting Public Participation   131

ment of personnel to train experts in this field, in particular for developing countries, and facilitate at the national level public awareness of, and public access to information on, climate change. Suitable modalities should be developed to implement these activities through the relevant bodies of the Convention, taking into account Article 6 of the Convention.

(Kyoto Protocol, 1997)

In 2002 states meeting at the Eighth Conference of the Parties to the UNFCCC agreed on the New Delhi Work Program, which aimed to integrate article 6 activities into existing climate change programs and strategies, promote synergies between conventions, and promote responses by intergovernmental organizations (IGOs) and NGOs. It listed a series of activities that states "could" develop to implement article 6. It is important to note that this is a nonbinding work program. Though the original time frame conceived was five years, it was extended for another five (i.e., until 2012) at the 13th Conference of the Parties to the UNFCCC, held in Bali in 2007.

In presenting its scope, the New Delhi Program (the version agreed to in 2002) redefined article 6's aims through separation and merging of different elements: "Parties are encouraged to undertake activities under the categories listed below, which reflect the six elements of Article 6: International cooperation...; Education...; Training...; Public awareness, public participation and public access to information" (UNFCCC, 2003, para. 10). Listing one element in isolation invests it with a stronger rhetorical force than placing it in a single paragraph with others. As we see it, public participation lost importance in this formulation of article 6. However, the program's 2007 amended version gives the same weight to all six elements and further specifies the modes of public participation: "It is useful to promote public participation in addressing climate change and its effects and in developing adequate responses, *by facilitating feedback, debate and partnership in climate change activities and in governance*" (UNFCCC, 2008; parts in italic are new to the 2007 version).

The New Delhi program makes important recommendations about the implementation of measures geared to promoting public participation. Paragraph 15 states that "Parties could ... : (d) Develop a directory of organizations and individuals, with an indication of their experience and expertise relevant to Article 6 activities, with a view to building active networks involved in the implementation of these activities." This could potentially widen the scope of civic involvement in climate politics.

The following part is especially valuable as far as public participation is concerned:

> Parties could ... : (i) Seek input and public participation, including participation by youth and other groups, in the formulation and implementation of efforts to address climate change and encourage the involvement and participation of representatives of all stakeholders and major groups in the climate change negotiation process.
>
> (UNFCCC, 2008, para. 15)[1]

The reference to "seeking input" from various groups of the public for policy processes and the specification of the stages of policy "formulation and implementation" are significant advances in the translation of the UNFCCC's article 6 goals. This enunciation appears to point to consultation of the public or even to more advanced forms of collaboration with citizens in the design and application of policy. The final part of the sentence refers to involvement in the "climate change negotiation process," presumably at the international level, but is a bit more limiting in terms of participants' identity: "representatives of all stakeholders and major groups." However, the following paragraph expands the scope of participant profiles:

> Parties should seek to enhance cooperation and coordination at international and regional levels, including the identification of partners and networks with other Parties, intergovernmental and non-governmental organizations, the private sector, state and local governments, and community-based organizations, and to

State Commitment to Promoting Public Participation   133

> promote and facilitate the exchange of information and material, and the sharing of experience and good practices.
> (UNFCCC, 2003, 2008)[2]

The UNFCCC determines that countries should regularly report the steps they are taking toward implementing commitments accepted under the convention (articles 4.1 and 12). Those reports have been designated national communications (NCs). The New Delhi Program also asks parties to report in their NCs "on their accomplishments, lessons learned, experiences gained, and remaining gaps and barriers observed" (UNFCCC, 2003).

The UNFCCC Secretariat provides guidelines for preparation of NCs. The first version of these guidelines was revised in 1999 and has continued to be applied to the latest NCs. That document constructs state obligations in relation to public participation in terms rather different from those related to education, training, and public awareness:

> Parties *shall* communicate information on their actions relating to education, training and public awareness. In this section, Parties *should* report ... on public information and education materials, resource or information centres, training programmes, and participation in international activities. Parties *may* report the extent of public participation in the preparation or domestic review of the national communication.
> (UNFCCC, 1999, para. 65; emphasis added)

The promotion of public participation is here demoted from a formal obligation to a voluntary act.

Public participation in environmental issues and in sustainable development has been consecrated in several other international agreements with which, as mentioned before, the New Delhi Program is intended to create synergies. For instance, in 1992 Agenda 21 proclaimed that "commitment and genuine involvement of all social groups" was "critical" to its implementation and that "one of the fundamental prerequisites for the achievement of sustainable development is broad public participation in decision-making" (Agenda 21, 1992). The Rio Declaration on Environment and Development determined that "environmental issues are best handled with the participation of all concerned citizens, at the relevant level" (Rio Declaration on Environment and Development, 1992). This was further established in the 1998 Aarhus Convention on Access to Information, Public Participation in Decision-making, and Access to Justice in Environmental Matters, which set up the right of the public to participate in decision-making processes.

The analysis of different legal and political documents suggests that although there are determinations for the promotion of public participation on climate politics, these are somewhat ambiguous. Whereas article 6 of the UNFCCC places public participation almost on a par with education, public awareness, training, and access to information, the Kyoto Protocol excludes any references to it. Even though some more recent documents diminish public participation in relation to the other issues, others detail states' commitments with regard to involving the public in climate politics.

RESEARCH DESIGN AND METHOD

We selected six countries for this project: China, India, Portugal, Tuvalu, the United Kingdom, and the United States. Criteria for country selection included UNFCCC status (half are Annex I countries to the convention, i.e., industrialized countries and economies in transition; half are non–Annex I countries, i.e., developing countries); contribution to global greenhouse gas emissions; vulnerability to climate change impacts; diversity of track records on climate policy; and potential representativeness of certain groups of countries, such as economies in transition.

As already mentioned, states are expected to regularly report progress on UNFCCC commitments via their NCs. According to the convention's website,

State Commitment to Promoting Public Participation   135

most of the 41 Annex I Parties submitted their first report ... in 1994 or 1995, their second in 1997–1998 and the third after 30 November 2001. The fourth NCs were due on 1 January 2006 and the fifth on 1 January 2010. Decision 9/CP.16 calls for submission of the sixth NCs on 1 January 2014.

(UNFCCC, 2012a)

Determinations are different for non–Annex I countries, which "shall submit [their] initial communication within three years of the entry into force of the Convention for that Party, or of the availability of financial resources (except for the least developed countries, who may do so at their discretion)" (UNFCCC, 2012b).

We collected the latest NCs available at the UNFCCC's website for the six countries at the time of our research (table 2).

Table 2. Latest available NCs for six countries.

Country	UNFCCC status	Number of current NC	Date of NC	Date of submission to UNFCCC	Number of pages
Portugal	Annex I	NC5	2010	30/12/09 (revised version 11/06/10)	248
United Kingdom	Annex I	NC5	2009	12/06/09	153
United States	Annex I	NC5	2010	28/05/10	180
China	Non-Annex I	NC1	2004	10/12/04	156
India	Non-Annex I	NC2	2012	04/05/12	310
Tuvalu	Non-Annex I	NC1	1999	30/10/99	38

Source. UNFCCC, 2012c. Data as of June 30, 2012.

We analyzed chapters in NCs related to the UNFCCC's article 6. Though we examined entire chapters, we paid special attention to references to the public, or citizens, and to public participation.

The analytical program that we employed draws on critical discourse analysis (e.g., Wodak & Meyer, 2009). The assumption here is that discursive practices, such as NCs (and more widely, political discourse), have a constitutive role: They create forms of intelligibility for the politics of climate change, and they produce effects in ways of thinking, speaking, and acting. The main points of the analytical framework are presented in the following paragraphs.

How is the relevant report chapter structured? What are the headings and subheadings? The document's organization into sections and subsections creates units of meaning (cf. Jäger & Maier, 2009) that matter in terms of the constitution of political issues and the definition of political priorities.

What is the order of topics in the text? Besides the semantic distinction and assignment of importance created by headings, the internal sequence of topics in each section also creates a hierarchical order. The superstructure (van Dijk, 1988) of NCs is briefly analyzed with a focus on the first paragraphs, which top-rank a given topic.

What lexical choices are made in reporting on the fulfillment of commitments? The concepts used by governments when speaking of (or avoiding) their responsibilities and achievements are associated with specific ways of thinking and acting upon them.

How is public participation discursively constructed? How is it defined and characterized?

How are the actors of public participation constructed? Who is this public? How is it defined, circumscribed, delimited, and selected? Is the diversity of publics acknowledged? In turn, how is the identity of the state defined? Is the state constructed as a promoter of public engagement, an authoritative source of knowledge or otherwise? In line with Halliday's (1985) systemic-functional theory of language, we posit that language not only generates representations of the world but also constitutes relations and identities. The system of (power) relations between different social groups is developed through discursive practices as well as material realities.

How does the state communicate with citizens? What forms and media of communication are employed to engage the public?

State Commitment to Promoting Public Participation 137

What vision of society and society–state relations is present? What other ideological aspects are present (explicitly or implicitly) in the text? Visions of public participation are likely to be associated with certain social, political, and cultural values and worldviews, and it is important to examine how these are discursively enacted.

Discursive practices always occur in a given social context, and the analysis of the historical conditions in which the reports were produced is likely to help one understand them (cf. Reisigl & Wodak, 2009). Therefore we briefly look into the following questions: What is the standing of each country in the international politics of climate change? What is its political situation? In the analysis, we also take into account nondiscursive social practices and (policy) materializations, as reported in the NCs.

AN ANALYSIS OF NATIONAL COMMUNICATIONS TO THE UNITED NATIONS FRAMEWORK CONVENTION ON CLIMATE CHANGE

This section examines the NCs of six countries to the UNFCCC and how they report on commitments these nations made under the convention's article 6. We start with an overview of relevant traits of the six countries and then move to a country-by-country analysis.

As argued earlier, contextual information can contribute to an understanding of the ways in which different countries may have responded to commitments regarding the promotion of public participation, as well as regarding education, public awareness, and training. Table 3 summarizes some relevant characteristics of the six countries covered in this chapter.

In what follows, we analyze each country's case. Countries are presented in order of their overall contribution to the global greenhouse effect in two groups (UNFCCC's Annex I and non–Annex I).

Table 3. Characteristics of the six countries analyzed.

UNFCCC status	Country	Type of government and political system	Population (July 2011 est.)	GDP per capita (2011 est.)	CO_2 emissions *	CO_2 emissions per capita**
Annex I	Portugal	Republic; Parliamentary democracy	10,781,459	$23,700	57,400	5.3860
	United Kingdom	Constitutional monarchy; Parliamentary democracy	63,047,162	$36,600	474,579	7.6977
	United States	Federal republic; Democracy	313,847,465	$49,000	5,299,563	17.2239
Non-Annex I	China	People's republic; Single-party system	1,343,239,923	$8,500	7,687,114	5.7585
	India	Federal republic; Democracy	1,205,073,612	$3,700	1,979,425	1.6389
	Tuvalu	Constitutional monarchy; Representative democracy (no political parties)	10,619	$3,400	Not available	Not available

Sources. U.S. Central Intelligence Agency (2012); United Nations Statistics Division (2012).
Notes. *Thousand metric tons (2009). **Metric tons of CO_2 per capita (2009).

Annex I countries

United States

Longtime holder of the number-one position in the world's greenhouse gas emissions table, the United States is currently the second largest emitter, producing about 18% of the world's annual total. It has one of the highest levels of emissions per capita. Although the United States signed the Kyoto Protocol to the UNFCCC, the country never ratified it. In 2001 George W. Bush declared that the country was abandoning the protocol because it was contrary to U.S. interests. In further international negotiations, the United States has opposed compulsory targets for greenhouse gas emissions without an equivalent commitment by developing countries. There is an ideological divide in the country regarding climate change; Democrats typically support action to address climate change, and Republicans generally oppose it. The central place of individual freedom in the national psyche is likely to come into conflict with a strong governmental role in responding to climate change.

The U.S. NC's chapter on "education, training and outreach" opens as follows:

> Federal agencies' climate change education, training, and outreach efforts seek to ensure that individuals and communities understand the essential principles of Earth's climate system and the impacts of climate change, and are able to evaluate and make informed and responsible decisions with regard to actions that may affect the climate.
> (United States Department of State, 2010, p. 139)

This sentence does two fundamental things. First, by speaking of "ensuring" that people "understand" the climate system and the impacts of climate change, the relationship between the state and citizens is here constructed as one of unilateral diffusion of knowledge. This excludes other, more dialogic forms of interaction. Second, the declared goal of ensuring that people are "able to evaluate and make informed and responsible decisions with regard to actions that may affect the climate" atomizes responsibility for addressing climate change and reduces the policy options for dealing with climate change to individual rather than societal or systemic options. Although attenuated by a reference to communities (in addition to individuals), the purpose is here presented as behavioral change with no mention of other forms of engagement, such as citizens collaborating in policy change.

The United States starts by defining the public target of education, training, and outreach as "individuals and communities." In presentations of programs that have been implemented, some references appear to communities (in some cases, professional communities), but it is unclear how community-level engagement has actually been put into practice (most programs that refer to communities also target individuals).

There are no references to public participation. This issue is all but obliterated from the state's agenda. There are mentions of "engagement," but the term is employed in a rather vague sense. For instance, in a section headed "Overview of National Efforts to Engage the United States on Climate Change," the NC mentions a disparity of developments in the previous years, from the publication of Al Gore's book and documentary *An Inconvenient Truth* and the organization of *Live Earth* to the release of the Fourth Assessment Report of the Intergovernmental Panel on Climate Change and to increases in media coverage of the issue. Later on, the subjects of engagement (and "participation") are specified, but what it entails is not:

> NOAA [National Oceanic and Atmospheric Administration] is committed to supporting and facilitating system-wide change of the formal education system to build educators' capacity to produce climate-literate citizens. Such change requires engagement and participation across the spectrum of the education community—including policymakers, academic institutions, professional associations, teachers, and students.
> (United States Department of State, 2010, p. 142)

Formal education "involve[s] K–12 and undergraduate curricula and postgraduate professional development programs." Whereas other countries analyzed here speak of "raising public awareness," the United States referred to "informal education programs" that have been "conducted in museums, parks, nature centers, zoos, and aquariums across the country" (United States Department of State, 2010, p. 140).

The United States' fifth NC describes a wide range of education and training activities. The emphasis is on the dissemination of information rather than on the promotion of transformations in practices and of institutions associated with the generation of greenhouse gases. Of over 100 federal climate change programs listed in the document, only one third

State Commitment to Promoting Public Participation   141

relate to action on greenhouse gas emissions, and the vast majority do so only from an educational point of view; the rest focus essentially on the detection of climate change and climate impacts. This appears to build on the assumption that more knowledge about climate change will lead to changes in behavior.

The United States used the term *outreach* instead of *public awareness*. Although the former is not employed by the UNFCCC, it may be more precise in designating what most governments appear to do: disseminate information rather than enact sustained policies geared toward the development of a conscious and responsive attitude.

Nevertheless, the following excerpt points to a different image, involving certain groups of people in the social and economic changes required to address climate change in a win-win formula:

> Education and workforce training are critical parts of EERE's [Department of Energy's Office of Energy Efficiency and Renewable Energy] mission, which is to create an energy-literate generation of skilled workers, leaders, and innovators who will produce affordable, abundant, and clean energy, thus accelerating the transition to a low-carbon economy and ensuring U.S. global competitiveness.
> (United States Department of State, 2010, p. 142)

The U.S. government referred to the need for "fostering public *climate literacy*—one that includes *economic and social considerations*" and argued that it "will play a vital role in knowledgeable planning, decision making, and governance" (p. 139; emphasis added). This is a positive advancement because it calls for the development of skills and competencies that integrate the biophysical sciences with society. In a unique move, the U.S. government stated the importance of a "comprehensive, interdisciplinary approach," speaking of integrating "the social sciences into federal agencies' educational and outreach programs" and maintaining that this "would help to ensure informed decision making and effective systems-level responses to climate change" (p. 139). However, in the NC the focus is scientific knowledge to the exclusion of other forms of knowledge, and the social sciences are mentioned in only a very few of the listed federal programs. The following excerpt also suggests that the U.S. government appraisal of the social sciences' role in addressing climate change is quite narrow: It appears to focus on a simplistic instrumental view of communication as information dissemination and to assume that scientific knowledge about the climate system leads to changes in individuals' decisions:

> The federal agencies are working with social scientists to *determine the most effective ways to communicate* with students and the public *about how Earth's climate is changing.* In an effort to extend their education and outreach programs and maximize their impact, federal agencies are addressing the following questions: How can local high-impact activities be scaled up and serve as national models? What are effective climate change literacy professional development opportunities for policy decision makers at all levels? How do we *assess changes in individuals' understanding of Earth's climate system* and the *decisions they make about their actions? How can nationally representative assessments of public knowledge and understanding of climate change help identify common knowledge gaps, misunderstandings, sources of confusion, and key concepts the American public needs to understand about climate change?*
> (United States Department of State, 2010, p. 140; emphasis added)

United Kingdom

The UK was the first country in the world to set up a legally binding, long-term framework for greenhouse gas emissions, the Climate Change Act (2008). It commits future governments to reducing CO_2 emissions by 80% from 1990 levels by 2050. However, implementation has been slow. In spite of some party differences, there is a political consensus in the country regarding the need to abate climate change. The UK has a long tradition of civic organization and civic action on environmental issues. A number of self-organized groups, such as Transition and

Low Carbon Communities, have been working toward the mitigation of climate change. A significant majority of the British public supports policies to combat climate change, although climate skepticism has grown in the last few years and there is resistance to some measures, such as the installation of wind farms.

The relevant chapter of the UK's NC is divided into two sections: education and public awareness. The educational focus seems to be on developing "the skills and knowledge" that children "need in a changing world," "an ability to *evaluate* environmental, scientific, and technological issues as well as *debate* informed, ethical views on complex issues, such as climate change," and encouraging "young people to *investigate, communicate* and *act* to tackle climate change" (Department of Energy and Climate Change, 2009, p. 128; emphasis added). In this case, education on climate change is depicted as involving more than the dissemination of knowledge, for the UK government highlights its aim of enabling and empowering young people to deal with climate change.

This NC includes multiple references to citizenship and to what being a citizen is about: "[young people's] role as global citizens," "the new curriculum ... has a focus on active citizenship"; "developing an understanding of environmental issues and how we lead sustainable lifestyles is a key element of becoming a responsible citizen" (Department of Energy and Climate Change, 2009, pp. 128–129). Good citizenship is defined as global, active, and green, and the state presents itself as a promoter of those virtues.

Public engagement is an objective associated with various activities described under the public awareness section. However, the identity of the public to be engaged is not always clear. For instance:

> ACT ON CO2, launched in 2007, is a major Government-led multimedia campaign ... which aims to *engage citizens* on climate change issues, address the confusion and powerlessness which can impede people taking action, and encourage genuine and sustained behaviour change to help reduce CO2 emissions.... The

> ACT ON CO2 website aims to signpost, interact, coordinate and *engage consumers* on climate change, providing a clear, consistent, authoritative and credible voice.... ACT ON CO2 calculator ... is an integral part of the Government's strategy to *engage with and educate the public* as part of the *mobilisation of society* to adopt low carbon lifestyles.
>
> (Department of Energy and Climate Change, 2009, p. 130; emphasis added)

This excerpt alone contains four labels with different connotations: citizens, consumers, public, and society. The rights, duties, and expectations of these different profiles are quite diverse, and the rationale for using one or the other is not apparent.

Despite these ambiguities, the UK government appears to be interested in learning more about different "segments" of the public:

> Defra [Department for Environment, Food and Rural Affairs] developed an environmental segmentation model, predominantly used for advising policy and communications development. It is based on people's responses to a broad range of attitudinal questions.... The model divides the public into seven clusters each sharing a distinct set of attitudes and beliefs towards the environment, environmental issues and behaviours. There has been a recent increase in the number of research projects and government bodies using the model. In addition, a web based tool, designed for use by a range of stakeholders to inform the public of which segment they best fit into and how they can make environmental changes, is currently in development.
>
> (Department of Energy and Climate Change, 2009, p. 130)

Although other publics are mentioned in the NC, the main focus of the British government appears to be young people. It is noteworthy that there is one reference to unions and business: "The Assembly Government [of Wales] is also looking at how we can develop a workplace-based component to the [communications] campaign [on climate change] and are interested in working with the Trade Unions and businesses on this." Business is also targeted by the Carbon Trust Campaign, which aims to help "businesses of all sizes" achieve "energy savings, reduce carbon emissions and make significant direct costs savings" (Department of Energy and Climate Change, 2009, p. 131).

There are no references to "public participation" in the report. However, the text points to various links between the state and "community action" in Wales:

> The Assembly Government is working to support community action on climate change. Activities have included:
> - Holding a series of community events to *find out more about what communities are doing* and *how we can support them better* as well as providing an *opportunity for learning, sharing experiences and networking*
> - Completing a scoping report on the action underway in communities and how this can be supported by the Assembly Government
> - Producing a Community Action Pack and associated DVD showcasing good practice which provides information on how communities and other groups can take action to tackle climate change.
> (Department of Energy and Climate Change, 2009, p. 131; emphasis added)

These are noteworthy forms of cooperation between the state and civic groups. The state thus positioned itself in a horizontal relationship with communities, as interested in learning about community initiatives and willing to support them.

Portugal
Under the Kyoto Protocol, the European Union (EU) committed to a reduction of 8% in its greenhouse gas emissions (in relation to the baseline year of 1990) but made a differential distribution of responsibilities among its member states. As an economy in transition, Portugal was allowed to increase its emissions by 27% but quickly went beyond this target, mainly owing to increases in road transportation. A strong invest-

ment in renewable energies in recent years has put the country back on track to meet its Kyoto target. Incongruously, the country continues to cut back on rail and other forms of public transportation. Portugal was ruled by a dictatorial regime for over four decades and became a democracy in 1974. It has a weak tradition of public participation and low levels of civic involvement with environmental issues.

In Portugal's NC, the chapter on education, training, and awareness raising includes four sections: (1) general policy guidelines on education, training, and public awareness, (2) primary, secondary, and higher education, (3) training, (4) raising public awareness, and (5) participation in international activities. It differs from others because it includes extensive information about the Portuguese school system and dedicates a short section to international cooperation (one of the commitments under UNFCCC's article 6).

Besides general subjects, such as "education for citizenship," Portugal's school system covers climate change in a few subjects up to grade 9 and in some tracks of secondary education. Moreover, the NC lists a number of school projects on environment and climate change. Most of the training activities for teachers have not focused specifically on climate change, instead addressing environmental issues and sustainability. Under "raising public awareness," the report lists a number of projects on energy and climate change aimed at the general public that have been led by a variety of agents, such as official agencies, local authorities, and corporations.

For the purposes of this chapter, the most relevant subsection of the NC is the one on "access to information and public participation." Portugal is the only country to include an explicit description of activities geared to public participation in its NC.[3] However, this is limited to public consultations: "The National Climate Change Programme (PNAC) and the National Allocation Plan [of greenhouse gas emissions allowances] (PNALE) were both subject to consultation processes" (Portuguese Environment Agency, 2010, p. 226). It is a posi-

State Commitment to Promoting Public Participation 147

tive development that the report includes information about how the process was conducted and what the outcome was: For "PNAC 2001, presented to the public on December 18th [2001], APA promoted three public sessions during the months of January and February 2002; results were included in the PNAC 2001 after analysis by the Commission for Climate Change" (Portuguese Environment Agency, 2010, p. 226). For "additional policies and measures" that were planned to meet commitments under the Kyoto Protocol, relevant "documents were disseminated and made available from APA website. Between December 18, 2003, and February 2004, comments and suggestions were received and summarized in the respective public discussion report" (Portuguese Environment Agency, 2010, p. 226). For PNALE, PNALE II, and the Portuguese National Strategy on Adaptation, the NC lists (or summarizes) those who participated in the public consultation. Significantly, for the first two, the vast majority are corporations and business associations with very little participation in so-called civil society. For instance, for PNALE,

> contributions were received from:
> 3 business associations
> 1 NGO
> 12 companies and other entities and
> 1 individual.
> All received individual replies from the working group, explaining and justifying the options adopted.
> (Portuguese Environment Agency, 2010, p. 226)

Though the NC report suggests that the government is open to reviewing its proposals and include contributions from all parties,[4] an analysis of the "public consultation report" relative to PNALE (Ministério da Economia and Ministério das Cidades, Ordenamento do Território e do Ambiente, 2004) reveals that six points coming out of the public consultation were incorporated in a new version of the document. They all corresponded to suggestions made by industry. None of the suggestions made by the only contributing organization representing NGOs—CPADA (Portuguese Confederation of Environmental

NGOs)—were incorporated, nor was their exclusion justified. Furthermore, although the NC creates an impression of accountability, CPADA complained about lack of transparency in the public consultation process because the greenhouse gas emissions of the businesses receiving emissions allowances were not made public (Ministério da Economia and Ministério das Cidades, Ordenamento do Território e do Ambiente, 2004, p. 9). It should be noted that Portugal was the only country to dedicate a subsection of its NC to the "involvement of non-governmental organisations" (with general data on state support for environmental NGOs; p. 228).

Confronting NCs with other documents and practices is not the aim of this chapter. Still, this minute exercise suggests that there may be a rhetorical function to the reports. Countries attempt to construct a self-image as open and dialogical. Other discursive and social practices, as well as policy materializations, may tell a different story.

Non–Annex I countries

China
Although its historical contributions to climate change have been small and its level of per capita emissions is still relatively low, China is currently responsible for the largest share of global greenhouse gas emissions (around a quarter). Despite significant investments in renewable energies, the country's greenhouse gas emissions are projected to rise sharply in the coming decades. This turns China into a vital player in the international politics of climate change, where it has tried to avoid any commitments based on its right to continuing economic development. As a single-party communist state, China's fortunes very much depend on official policy decisions. However, given the sheer size of its population and the speed of its socioeconomic transformation, citizens' attitudes and behaviors toward climate change are likely to increasingly make a difference.

In its NC, China clearly connected climate change with sustainable development. The opening paragraph of the chapter on "education, training and public awareness" reads as follows:

> In the *Program of Action for Sustainable Development in China in the Early 21st Century* formulated by the Chinese government in 2002, it was put forward: to develop education at all levels and in various ways and enhance public awareness of sustainable development, and to reinforce human resource development to build up the public's scientific and educational capacities to participate in sustainable development.... China has ... devoted considerable efforts to raise public awareness on climate change to promote sustainable development.
> (People's Republic of China, 2004, p. 139)

An emphasis on climate change as an "issue of development" is also found in China's National Climate Change Program (Chinese National Development and Reform Commission, 2007). Placing climate change within the economic and social framework of development has been a key aspect of the position of China, as well as those of countries like India and Brazil, in the international negotiations on climate change. The entitlement to development has been used as a central argument in these countries' rejection of commitments to reduce greenhouse gas emissions. "As this is, in many ways, a *moral* argument, it is all the more powerful in the legitimation of the position of China and other countries in the international politics of climate change" (Carvalho, 2008, p. 6).

China's NC is organized under three main headings: "education and public awareness raising," "training and public awareness raising," and "outlook." Although there are specific references to climate change, education and training on climate change appear mainly subsumed under environmental education: "China has already included environmental education in the Outline of All Subjects for Compulsory Education in Primary and Secondary Schools ... training for officials in the environmental protection sector has been greatly strengthened" (People's Republic of China, 2004, p. 139).

The section on training and public awareness is divided into multiple subsections highlighting different means and mechanisms of dissemination of information on climate change. After presenting survey data that point to the centrality of media (both television and newspapers) for the public, the Chinese government included a heading on media publicity, under which it is argued that "China has made full use of the media including TV, broadcasting and newspapers for the publicity of environmental protection and climate change" (People's Republic of China, 2004, p. 141). This is surprising, for research shows that references to climate change in some Chinese media were very sparse at the time China's NC report was published. That is at least the case of *China Daily*, an English-language newspaper that published 53 articles in 2004 containing the phrase *climate change*; the number jumped to 635 in 2007 (Carvalho, 2008).

The report also includes subheadings on website construction, public lectures and reports, workshops and forums, and publications and other training materials. The Chinese government portrays itself as an active agent in the promotion of public awareness and the dissemination of knowledge on climate change through the Internet, publications, and other media. It also suggests that it engages in some form of dialogic communication with its citizens: "China has also used the Internet to conduct experts' lectures and organize the experts to exchange online with the public on the questions of climate change" (People's Republic of China, 2004, p. 142).

In China's NC, the emphasis is on public education for adaptation to climate change—"surviving education," "precaution education"—not for mitigation. Mitigation does get a mention in a defensive way: "It is necessary to raise the awareness of the business managers and staffs, making enterprises to become aware of the pressures in mitigating climate change and the underlining relationship between the counter-measures and the development of enterprises" (People's Republic of China, 2004, p. 146).

State Commitment to Promoting Public Participation   151

There is only one explicit reference to public participation in China's NC: "The whole society has witnessed increasingly active development in training on sustainable development, environmental protection and climate change with a higher public participation" (People's Republic of China, 2004, p. 141). What *public participation* means here is unclear.

India
The world's largest democracy is a complex social terrain in which fast economic growth and high-tech leadership coexist with widespread poverty and a dire lack of basic infrastructure. With rather low levels of per capita greenhouse gas emissions, India has objected to making any pledges regarding its future emissions; it has become one of the leaders of the developing world in international negotiations, recurrently defending "equity" rights in greenhouse gas emissions. India stands to suffer severe impacts from climate change, such as cyclones and storms, heat waves, reduced water availability, drops in agricultural yield, and forced coastal displacements.

The theme structure of India's NC focuses on the scale of action, as well as on its agents: activities at the national level and at the state level, government-supported and private sector initiatives. In a lengthy chapter, a wide number of activities and initiatives are described, many of which bear no connection to education, training, or public awareness touted in the chapter's title.

From the outset, India's NC constructs awareness of climate change in clear association with scientific knowledge. The following are the document's opening lines:

> In line with the Government of India's commitment to spreading awareness about climate change education and strengthening the scientific network, the National Mission on Strategic Knowledge for Climate Change (NMSKCC) was identified to build a vibrant and dynamic knowledge system that would inform and support

> national action for responding effectively to the objectives of sustainable development.
>
> (Ministry of Environment and Forests, 2012, p. 223)

In the first paragraph, the dissemination of knowledge on climate change is connected to economic and social development:

> The National Knowledge Commission to the Prime Minister was also identified as being an important component regarding climate change education.... [T]he Commission would carry out the following activities.
> - Build excellence into the educational system to meet the knowledge needs/challenges and increase India's competitive advantage....
> - Promote knowledge applications in agriculture and industry, and knowledge capabilities to make government an effective, transparent, and accountable service provider to the citizen.
> - Promote widespread sharing of knowledge to maximize public benefit.
>
> (Ministry of Environment and Forests, 2012, p. 223)

The NC offers a sharp analysis of India's National Action Plan on Climate Change, which (like China's) prioritizes economic development over climate-related policies:

> The Plan ... starts by first and foremost marrying climate change to development concerns in no uncertain terms. The very first line states, "India is faced with the challenge of sustaining its rapid economic growth while dealing with the global threat of climate change." Thus, the goal of development is unambiguously underscored, and climate change is recognized as a major problem, not least because it could hurt development targets.
>
> (Ministry of Environment and Forests, 2012, p. 224)

Much of the chapter focuses on expert knowledge and appears to refer to elite, specialized publics rather than to the general public. For instance, the Environmental Information System is said to be aimed at "decisionmakers, policy planners, scientists, engineers, and research

State Commitment to Promoting Public Participation 153

workers" (Ministry of Environment and Forests, 2012, p. 224). Nonetheless, there is a strong (rhetorical) emphasis on the role of environmental information for the whole of Indian society, with activities targeting "students, youths, teachers, tribals, farmers, other rural population, professionals, and the general public" (i.e., the National Environment Awareness Campaign; Ministry of Environment and Forests, 2012, p. 234). A wide range of activities and media is employed to that purpose: folk dances and songs, street theater, puppet shows, films, television programs, workshops, mobile exhibition vans, and so on. It must be noted that although many activities are about climate change, the majority of them concern other environmental issues or involve general environmental education.

Reflecting the nature of India's socioeconomic system, many activities relate to agriculture and basic necessities, such as water management and sanitation. Initiatives often involve multiple goals, including responding to fundamental needs:

> A Knowledge based System ... and the project "Mobilizing Mass Media Support for Sharing Agro-information" [are] expected to provide crucial information for accelerated and sustainable transformation of Indian agriculture through print and electronic mode, targeting Panchayati Raj institutions, private sectors, and other stakeholders. It is envisaged that such an intervention would help in poverty alleviation and income generation.
> (Ministry of Environment and Forests, 2012, p. 228)

Despite occasional mentions of mitigation, most attention seems to go toward adaptation to climate change—for example, "all the state governments were called upon to prepare State Level Action Plans on Climate Change (SLAPCC)[, which] will enable communities and ecosystems to adapt to the impacts of climate change effectively" (Ministry of Environment and Forests, 2012, p. 230).

Both at the national level and at the state level (with large differences in levels of investment between states), India's NC refers to a number of other agents of education, training, and awareness raising: "a network of nodal agencies and grassroot-level organizations," as well as "NGOs, schools, colleges, universities, research institutes, women and youth organisations, army units" (Ministry of Environment and Forests, 2012, p. 233), private companies, and other private-sector organizations (e.g., through corporate social responsibility projects) and foreign bodies (e.g., USAID). Some of the nonprofit organizations have an international nature or a link to another country (e.g., the Climate Project, founded by Al Gore).

Whereas the Indian government made some statements in its NC regarding the importance of public participation in responses to climate change, it appears to leave actual engagement initiatives to NGOs. The following paragraphs discuss two of the most significant projects that the government claims to support.

The M. S. Swaminathan Research Foundation, a nonprofit trust promoting development and employment of poor women in rural areas, is said to have

> hosted an inter-disciplinary dialogue on the theme "Community Management of Climate Change: Role of Panchayats and Nagarpalikas" to prepare a well-defined roadmap for empowering local communities with knowledge and skills relevant to enhancing their capacity to manage the adverse impact of climate change. A series of consultations involving various panchayat leaders have been initiated to discuss the possible components of such a legislation. [It also hosted] a "National Dialogue on Adaptation to Climate Change." Participants included Cabinet Ministers, Prime Minister's Special Envoy for Climate Change, Secretaries from various ministries, State Secretaries, donor agencies, members of Planning Commission, academics, and various NGOs.
> (Ministry of Environment and Forests, 2012, pp. 236–237)

The Centre for Social Markets, a nonprofit organization "harnessing the power and potential of markets, entrepreneurs and other economic actors to do good" (Centre for Social Markets, 2012, n.p.)

> runs the Climate Challenge India initiative to reframe the climate debate in India and to create a proactive, opportunity-led approach towards addressing it.... Public awareness building and mobilization to make political constituencies more receptive to the need for change are central to the campaign. Business and city elites are another target for focused engagement and leadership.... Working closely with the arts and culture community, the [City Dialogues on Climate Change] campaign uses creative media technology to reach out to India's geographically and linguistically diverse communities, building a nationally relevant knowledge and communications platform in the process.
> (Ministry of Environment and Forests, 2012, p. 237)

These activities deserve to be recognized for the diversity of publics reached, including disadvantaged groups, and for the methods and media of engagement. However, in a country where many aspects of state responsibility have been transferred to NGOs and thus removed from public scrutiny and public choice, that the promotion of public participation on climate politics is also dislodged from the realm of the state should be a matter of concern.

Tuvalu

Tuvalu is a small Pacific nation comprising nine islands. As a low-lying and least developed country, it is quite vulnerable to sea-level rise, storms, and other impacts of climate change. Tuvalu is a member of the Alliance of Small Island States (AOSIS), an IGO that was formed to strengthen its members' voices in international negotiations on climate change. AOSIS has played a very active role in the international politics of climate change, and Tuvalu has sternly defended the limitation of the global average temperature rise to 1.5 degrees Celsius rather than the 2 degrees Celsius agreed on by most parties to the UNFCCC. No official data is available for Tuvalu's greenhouse gas emissions, which are likely to be minute.

As of July 2012, Tuvalu had submitted only one NC to the UNFCCC. It is dated 1999. The document is extremely short, at a total of 38 pages. References to the areas under article 6 are very brief. Nonetheless, the country reported on progress in education and awareness-raising activities, which are considered a "priority" (Ministry of Natural Resources and Environment, 1999, p. 19). Unlike other countries (and particularly developing ones), Tuvalu referred in its report to program measures focusing specifically on climate change and not on the environment as a whole (or on sustainable development). The country's NC states that climate change has been incorporated into primary- and secondary-school curricula, and the University of the South Pacific has a postgraduate course on vulnerability and adaptation.

Several activities listed in the following excerpt appear to call for active citizen engagement:

> In terms of public awareness, a strong position has been taken by the government and communities through *participatory radio programmes* (interviews), leaflet production, *essay competitions*, *poster competitions*, national workshops and visits to outer islands to promote education and awareness on climate change and sea level rise.
> (Ministry of Natural Resources and Environment, 1999, p. 19; emphasis added)

Presenting its "future directions," the Tuvalu government mentioned the "appointment and training of a dedicated public educator who would work with the Department of Education in schools, the media and with the public to improve cultural attitudes to the environment and clarify misconceptions" (Ministry of Natural Resources and Environment, 1999, p. 23). This suggests a tendency for centralization and cross-sector integration of measures to promote public awareness, a tendency that may not be surprising in a country of tiny size and resources. Although no

State Commitment to Promoting Public Participation 157

data are given in the document, the UNFCCC has reported very low levels of awareness in many developing countries and has called for funding of article 6 measures through the Global Environmental Facility (UNFCCC, n.d.).

CONCLUSIONS

Substantial transformations in social and political practices would be required to mitigate climate change and avert its worst impacts. Policies and forms of governance for climate change and all concomitant issues (energy production and consumption, industrial development, transportation, etc.) would have to be modified. Enacting the kind of political change that is required to deal with climate change calls for wide citizen engagement; public participation in policy processes, despite its limitations, is likely to generate positive outcomes.

In this chapter, we have analyzed international commitments to promote public participation in the development of climate change policy and the ways that six countries have reported on the fulfillment of those commitments. We detected tensions and ambiguities in the legal and political documents that define state responsibilities. The UNFCCC builds public participation explicitly into states' commitments but lessens it vis-à-vis education, training, and public awareness, the only issues that are placed in the title of its article 6. The Kyoto Protocol excludes any references to public participation. Some of the official guidelines for producing NCs to the UNFCCC also appear to erase the promotion of public participation from the realm of state responsibility, constructing it as an option rather than an obligation. Still, the New Delhi Work Program proposes that states seek input and public participation in the formulation and implementation of climate policy.

Analysis of the NCs of six countries suggests that governments have been making the UNFCCC's commitments on public participation a dead letter. Most countries do not refer explicitly to public participation or else make only very vague statements about it. Portugal, whose NC includes a subsection on public participation, is an exception. However, this country referred only to public consultation, which is a minimum form of participation, or tokenism by Arnstein's (1969) standards, and appears to fulfill rhetorical goals instead of reflecting political intent. In any case, it would be useful to know who participates in these exercises and what their impact is in other countries, such as the UK, which has also put in place public consultations on policy proposals for climate change but chose not to refer to these processes in its NC. The UK report can nevertheless be singled out for the Welsh government's acknowledgment of the importance of activities and initiatives on climate change led by communities, and for its declared intent to learn about them.

Whereas the UNFCCC's determinations in terms of state-sponsored public participation in developing appropriate responses to climate change would call for the promotion and proactive facilitation of citizens involvement in policy processes, most governments excluded these commitments completely from their reports and, hence, from their agendas. Their overwhelming silence on this matter indicates a widespread intent to turn public participation into a political nonissue.

In a report on the "role of public engagement in climate change policy" produced for the UK's Sustainable Development Commission, Creasy, Gavelin, Fisher, Holmes, & Desai (2007, p. 10) referred to *political space* as

> any public debate in which government representatives, either political or administrative, are called to interact with an issue and respond on behalf of government." Political space can lead to policy proposals or it may cause governments to defend the status quo, reflecting how often agendas for public policy can be contested by differing actors in the policy-making process. Political space activity can be said to become part of the policy direction stage of the policy-making process when it leads to actual action.

State Commitment to Promoting Public Participation 159

Judging from the official reports of six countries, it appears that governments do not create such political spaces on climate change. Although this was not to be expected of political regimes that admit to an authoritarian administrative structure, nominally democratic countries arguably should create mechanisms and forms of dialogue with their citizens on this all-encompassing issue.

The relationship between state and citizens is generally constructed in the NCs in vertical terms; the state positions itself as the source of knowledge to be disseminated to the public, which is construed as a passive recipient. This unilateral diffusion of knowledge conforms to the information-deficit perspective, and the concept of public engagement with science and technology is nearly universally absent from the reports.

With minimal exceptions, wherever skills to evaluate, debate, and act upon (information about) climate change are mentioned, the acquisition of *scientific* knowledge by the public appears to be the state's goal in both industrialized and developing countries. Given the ways governments discursively construct these matters, they seem to build on the assumption that more knowledge about climate change will lead to changes in behavior, despite research showing that this is not the case (e.g., Carvalho, 2011). Perhaps not surprisingly, developing (non–Annex I) countries place climate change in the context of the right to economic development. Education and awareness raising are directed to adaptation, not mitigation, which is viewed as the responsibility of others.

In most cases, the public is a relatively vague entity. Sometimes particular groups are singled out, but except for young people, who are the targets of formal education, there is no sustained description of how groups are reached. Bora and Hausendorf (n.d., p. 2) viewed citizenship as "an ongoing communicative achievement of social categories rather than a mere result of civil rights and entitlements that actors are supplied with." The communicative value of these reports lies in the fact that through them, states construct citizens as subjects to be educated and governed without any political capacity. Reminiscent of ideas on "governmentality" (Dean, 1999), the visions of state–society relations enacted here appear to rely on formal and informal education as techniques of control while rejecting civic participation in policy processes.

Endnotes

1. The original and the amended versions of the program are identical regarding cited text, (d) and (i).
2. This paragraph is also found in both versions of the program.
3. Portugal's NC is also the only one that makes connections to other relevant conventions (as suggested by the New Delhi Working Program): "In January 2005 APA [Portuguese Environment Agency] published the 1st National Report on the Aarhus Convention, on Access to Information, Public Participation in Decision-making and Access to Justice in Environmental Matters. This Report refers to measures adopted to assure that the MAOT [Ministry for Environment and Spatial Planning] bodies and their employees support and assist the general public" (Portuguese Environment Agency, 2009, p. 226).
4. For instance, the current "version of the proposal for the [National] Strategy [on Adaptation] has been sharpened to include the comments and suggestions received and it has been proposed for formal approval" (Portuguese Environment Agency, 2009, p. 227).

CHAPTER 6

CLIMATE GOVERNANCE AND VIRTUAL PUBLIC SPHERES

Anna Maria Jönsson

> Through Second Life, it doesn't matter if you're a farmer in Turkmenistan or a professor at Harvard. Everyone can ask questions anonymously and everyone's opinion is just as valid.
> (Allen, 2007, n.p.)

ENVIRONMENTAL GOVERNANCE AND THE ROLE OF THE MEDIA

How societies should come to terms with pressing environmental problems, such as climate change, is one of the most debated questions in contemporary public discourse. Modernity has brought an increased production of risk as well as new kinds of risks that societies need to deal with. Nowadays, risks are often global and cross physical and cultural boundaries (Beck, 1992). Through societal changes like globalization and deregulation, politics and decision-making processes face new challenges: Transboundary risks have raised a need for new political spaces and for a redefinition of political identities (see, e.g., Castells, 2008).

The concept of *environmental governance* focuses on new ways of decision making and regulation in a world that is increasingly complex, multilayered, globalized, and less state-centric (cf. Joas, Kern, & Sandberg, 2007). Various social actors (policy makers, social scientists, etc.) seem to agree that for societies to manage and govern global risks, there is a need for transnational communication and decisions, as well as for multistakeholder participation and increased citizen involvement. This is sometimes described as "good governance" (Whiteside, 2006). Good governance includes aspects like transparency and participation, and as a normative model it shares several characteristics with the notion of deliberative democracy and the public sphere ideal (Habermas, 1989). There are, of course, numerous models of democracy in which public participation is not included, but the deliberation model is certainly the one with the strongest links to the theory and practice of environmental governance and also with the ecological-symbolic perspective (cf. Zavetoviski, Shulman, & Schlosberg, 2006). According to Carpentier (2011), in models of deliberative democracy, participation is generally understood as communication. All these perspectives focus on the importance of an open and inclusive communicative space and what Dryzek (1994) called a "communicative rationality" (or discursive democracy).

According to Peter Dahlgren (2006), the notion of deliberative democracy has come to equal the model for communicative interaction in the public sphere. However, he argued that the notion of deliberative democracy needs to be problematized because it ignores several important aspects in relation to citizenship and civic competence. Dahlgren discussed the limitations of deliberative democracy and whether there is a need to rethink the role of citizenship or what one means by *the political*. Nonpolitical contexts of civil society and everyday life experiences are, for example, also likely to have a bearing on how people engage in political discussions. Through communicative practices and interaction, citizens can constitute themselves as "imagined communities" (Dahlgren, 2006). Here the media have an important role to play.

Modern democracies are generally characterized by mediated participation, in which media act as mediators between policy makers, experts, and the public (cf. Dahlgren, 1995). Following the work of the German sociologist Jürgen Habermas (1989), a considerable number of researchers have used the concept of the public sphere to describe and to evaluate the role of the media (e.g., Örnebring & Jönsson, 2004), and one conclusion is that media can shape public discourse in terms of participation and representation. The media also play a crucial role in defining what constitutes political problems and in framing issues such as environmental risks (Allan, Adam, & Carter, 2000; Anderson, 1997; Cox, 2006; Hansen, 2010).

The public sphere is not by any means unitary but instead consists of different communicative spaces, often described as alternative public spheres or counter public spheres (cf. Dahlgren, 2006; Fraser, 1992), and whereas the so-called mainstream public sphere is characterized by the domination of elite voices, alternative public spheres provide space for the regular citizen and different kinds of minorities. Yang and Calhoun (2007), for example, presented the idea of a green public sphere, a public sphere with particular relevance for the area of environmental governance. According to their definition, a green public sphere is a space for fostering political debates and pluralistic views about environmental issues. It can be said to consist of mainly three elements: an environmental discourse, publics that produce or consume this discourse, and media used for producing and consuming the discourse (Yang & Calhoun, 2007). The discourse is normative and strives toward engagement for a sustainable society, meaning, among other things, a sustainable environment. It can be noted that a number of studies focus on ways that new social movements, such as (environmental) activist groups, communicate and use what is often called "alternative media" and new media as platforms (see e.g., Bennet, 2003; DeLuca, 1999; Taylor, Kent, & White, 2001).

Although governing and public decision making have traditionally been closely linked to the nation-state, environmental risks like climate change need to be deliberated and governed on a much broader scale, involving actors at several different levels and in different parts of the world (cf. Renn, 2008). Several studies on risk governance underline the importance of communication and extended participation of stakeholders (such as nongovernmental organizations [NGOs]) and the wider public (Klinke & Renn, 2002; Stirling, 2006). It is fair to assume that this situation could generate a need for new, alternative public sphere(s) because the mainstream public sphere is often considered to be dominated by the power elite. This chapter discusses whether online media and the new communicative spaces that they allow for can function as alternative public spheres or whether they are simply an extension of the mainstream traditional public sphere.

GLOBAL SOCIETY, DELIBERATION, AND ONLINE PUBLIC SPHERES

Like many other scholars, Manuel Castells (2008) highlighted the importance of analyzing and understanding globalization and claimed that this process has led to a shift in the debate's focus from the national to the global domain. This has led, among other things, to the emergence of a global civil society and to ad hoc forms of global governance and civic agency. As already discussed, one of the main ingredients in the public sphere is communication and interaction among citizens, civil society, and the state. According to Castells (2008), the public sphere as a space of debate on public matters has also shifted from the national to the global arena.

In the globalized network society, the public sphere is organized more than ever on the basis of media communication networks, including mass media, the Internet, and wireless communication (cf. Bennett & Entman, 2001; Castells, 2008). "That global public sphere is built around the media communication system and Internet networks, particularly in the social spaces of the Web 2.0, as exemplified by YouTube, MySpace, Facebook, and the growing blogosphere" (Castells, 2008, p. 90). These spaces for communication are sometimes labeled "online public spheres" (Benkler, 2006; Dahlgren, 2009; Gripsrud, 2009). The public sphere here is seen as a facilitator of communicative exchange and public discourse that claims universal validity and goes beyond representation of certain interests, values, or traditions (cf. Trenz, 2009).

Several studies have focused specifically on the political sphere and on whether and how digital media can be understood in terms of power and democracy (see, e.g., Dahlberg, 2001a; Dahlberg, 2001b), as well as whether Internet communication can work as an extension of the public sphere (Dahlberg, 2001b). Issues like the inclusive and deliberative potential of digital media, especially the Internet, and how different forms of user-generated content are linked to citizenship and participation (see, e.g., Carpentier, 2011; Dahlgren, 2011) have been widely discussed. Dahlberg (2001a), for example, compared deliberative practices in online media with the Habermasian normative model of the public sphere. His analysis shows that rational critique and exchange of positions do take place within many online spaces, but that there are different factors limiting the expansion of the public sphere online. These factors include, for instance, the dominating role of state and corporate interests, a reflexivity deficit, the exclusion of the wider public (the "many") from these online political spaces, and the fact that the discourse is dominated by certain individuals and groups.

Political Discourses and Web 2.0

Discourse can be broadly defined as use of language and a particular way of representing some part of the world (see, e.g., Fairclough, 2003). In this chapter the concept is understood and defined at a general level as discussion in conversational communities (cf. Berglez, 2010). It has been shown that different media can reshape and adapt to political discourse in different ways, and it is fair to assume that a changing media landscape also contributes to a change in the conditions for politics.

Four arenas, all part of what is termed Web 2.0, have been identified as constituting the foundation for the virtual society: blogs, social networks (e.g., Facebook, YouTube, Flickr, MySpace), virtual worlds (e.g., Second Life), and collaborative systems like Wikipedia (Wahlström, 2007). In different ways, these arenas represent new forms of networked public-sphere activities in which citizens and civil society actors play the main role. Certainly, blogs and collaborative systems can also function as deliberative tools, but in many respects they still represent the traditional sender-focused kind of mediated communication. There is ambivalence in academia regarding the democratic potential of online media, and whereas some point to their inclusive nature and deliberative functions, others claim that they simply reproduce existing inequalities in terms of power and participation (cf. Dahlberg, 2001b; Zavestoski, Shulman, & Schlosberg, 2006) and that there is a clear "political economy" in online media (Jönsson & Örnebring, 2011; Mansell, 2004). Both perspectives, though, underline the *potential* of online media as deliberative forums for a discursive democracy.

Much attention in academia has also been directed toward the issue of how media affect politics (see, e.g., Mazzoleni & Schultz, 1999). For a long time television was the focus, but since the mid-1990s the Internet has become the center of interest. Thus researchers in media and communication studies and in political science have directed quite extensive interest toward the issue of how the Internet and online social media, such as Facebook and Twitter, have affected political discourse and the relationship between politicians and citizens (see, e.g., Klotz, 2001; Small, 2008), as well as their role in conflicts and revolts (see, e.g., Burns & Eltham, 2009). Previous research on the role of the Internet and Web 2.0 in political communication has shown that even if mainstream news sources still dominate how voters get information online about the presidential election campaign, a majority of users now also get information from blogs and various websites (Gurevitch et al., 2009). Not surprisingly, young people are heavily represented among these new-media users.

In order to respond to the buzz of media interactivity many politicians, NGOs and other actors use blogs, Facebook, YouTube, and others to communicate (Gurevitch, Coleman, & Blumler, 2009). It has been argued that online communication can be expected to affect collective action and may have different implications for different kinds of organizations (see, e.g., Diani, 2000), social movements, and networks. Bennett (2003) has studied how global activist networks use digital media like the Internet to plan, communicate, and coordinate their activities. For these activist networks, digital media work both as a means to communicate internally and as a means to make their messages known to the wider public. The Internet also contributes in many ways to global public images of events, as in the case of the World Trade Organization meeting in Seattle in 1999, and works as a way to tie activists together in virtual political spaces (Bennett, 2003).

Environmental (political) discourse(s) is to a large extent dominated by the science–policy interface, whereas the public space for citizens' communication has been rather limited. The Internet has created new communicative networks and allowed citizens to take part in the production of content and to communicate with friends and strangers all over the world. Research on Web 2.0 and so-called social media (e.g., Facebook, MySpace) has focused on issues of identity, integrity, interpersonal relationships, and so on, as well as on their importance for political communication and the public sphere (see, e.g., Dahlgren, 2005; Fornäs, Klein, Ladendorf, Sundén, & Sveningsson, 2002; Turkle, 1995), but less attention has been directed toward the question whether and how these networks can function as public spheres and platforms for political deliberation in the particular area of governing environmental risks, such as climate change (nevertheless, the field has attracted increased interest the last couple of years). These issues are addressed in this chapter.

How environmental risks are framed and perceived is of utmost importance for how societies are governed. Framing theory is well suited for analyzing how actors can be involved in public discourse, debating, defining, and setting a particular agenda and furthering its implementation (Rein & Schön, 1993). In studies of the role of media and journalism in public discourse, framing is generally used as a way to describe ways that different issues are represented in the news. The concept of framing can be said to consist of two main dimensions: The first is how certain aspects in a communicative text are made more salient, and the second, how these frames affect the way people perceive reality, societal problems, risks, and so on (Entman, 1993). Here I address these issues through a study of environmental discourses in different online media with a focus on climate change.

Case studies

I analyze how communicative spaces like virtual Internet worlds and social media can be used as platforms for communication on environmental issues, such as climate change, from two perspectives: The first focuses on individuals and how user-generated content in different online media—Facebook and Second Life—frames the issue of climate change; the second perspective focuses on how organizations engaged in environmental governance and discourse—here represented by Greenpeace as a global NGO—make use of these online spaces for communication. I also discuss whether and how these platforms can be understood as (green, alternative) public spheres. The analysis presented here is exploratory but tries to identify general criteria concerning the role of online media in environmental communication.

Facebook and climate change discourse

Facebook is a globally spread website for social networking founded by Mark Zuckerberg and launched in 2004. Its recorded growth has been very rapid, but it is difficult to find reliable information about the exact, up-to-date number of users. According to the Associated Press ("Number of Active Users," 2012), the number of active users was 901 million in March 2012. Because of the site's popularity, research on Facebook has increased rapidly, focusing on issues like social interaction, trust and privacy, identity, and also the role of Facebook (and other social media) in political communication (see, e.g., Alexandrova, 2011; Jackson & Lilleker, 2009; Westling, 2007; Williams & Gulati, 2008). Research on Facebook in relation to environmental issues is rather limited. Bortree and Seltzer's (2009) study on Facebook profiles of environmental advocacy groups suggests that the use of dialogic strategies to create opportunities for dialogic engagement can lead to an increase in the number of stakeholders who interact with the groups and organizations.

Facebook is mainly used as a tool for social networking and communication in the private sphere. It can also be used by high-profile celebrities and elite individuals to disseminate their messages and to develop dialogic exchanges with audiences or citizens. In addition, Facebook serves as a tool for creating communities (based on a common interest or opinion, for example) and is also used by existing communities to communicate among themselves and with others. One application on Facebook, in fact, allows users to create and join different groups. Groups are countless and cover various subjects in both the private and the public sphere. Some groups have only one member, whereas others boast hundreds of thousands. This application is one illustrative example of how Facebook can be used as a forum for public discourse and environmental communication. This case study examines how climate change is framed in Facebook groups.

To find relevant material, I searched Facebook for groups using the key phrases *climate change* and *global warming*. This yielded over 4000 groups. To gain a broad overview of the material, I then categorized the first 500 of these groups according to type or category and number of members. From among the first 100 groups in the search result, I selected 50 for further analysis of the framing of climate change issues. I chose those 50 groups to represent different sizes, level of activity, parts of the world; they include both groups promoting the combat of climate change and skeptics' groups. Actors and activities in the (user-generated) content in these groups were analyzed, as were the sources of information and representation of (scientific) knowledge, and the ways the groups presented climate change in terms of problems and solutions. Elements like the title of each group and its use of different symbols and images, together with the participants' and the group administrator(s)' contributions, are other important parts of the framing process. The process of framing is interactive, but the degree and form of interaction vary between different contexts and different arenas (Mormont & Dasnoy, 1995).

To understand how climate change is framed in Facebook through activities in different groups, it can be illuminating to cast an eye at how the groups characterize themselves. Among the groups on climate change, the most common categories (in 2010) were "student groups," "organizations," and "interest groups." Together these categories represent slightly less than 50% of all the climate change groups. Many groups were also categorized under "corporations" and "geography," as well as "entertainment and culture" and "just for fun." Least common were the categories "music" and "sports and leisure." The groups have between one and 400,000 members (most have between 20 and 3,000). They differ greatly in activity level and problem definition, and there are groups both with and outside ongoing discussions. For some examples see table 4, in which three groups have been selected to represent different group types.

Table 4. Three examples of climate change groups in Facebook, September 2010.

Name	Main focus, Self-presentation	Status of activities
Join the Fight to Stop Global Warming!	Raise awareness, consumer behavior "A big part of global warming is the by product of manufacturing and waste from the things we use and consume. We need to live smarter. We need to reduce reuse and recycle!" (Join the Fight to Stop Global Warming!, 2010)	Very active
100 Million Voices for a Real Climate Deal in Mexico 2010	To create pressure on the way to Climate Summit Mexico 2010 "This group was founded in the utter disappointment brought about by the shameful outcome of the 2009 U.N. Copenhagen Climate Conference (COP15). An overwhelming number of people have already joined, and our numbers are growing daily as people have found not only a place where they are able to show their support for a real and effective climate change deal, but also a place where they are able to express their opinions and exchange their ideas on how we can bring about the necessary change towards a more sustainable economy and way of life." (100 Million Voices for a Real Climate Deal in Mexico 2010, 2010)	Very active
Global Warming??	Skeptics. Work against the idea of global warming as a threat. "So, all you hear lately is how we humans are killing planet earth with our CO2 (Carbon Dioxide) gasses and if we don't do something now, our planet is DOOMED! Well this group is for anyone and everyone who believe that: 1.) Global Warming is NOT A THREAT to our planet. 2.) Yes, our planet's temperature has risen...and fallen in the past decade/century/ or even its existence...IT is normal and humans do not drive climate/global temperature change. It is natural for global temperatures to fluctuate. 3.) The Global Warming hype is largely being used to pass legislation that will tax 'unfriendly environmental machines.' 4.) Going Green and reducing pollution is great! However, we know that Carbon Dioxide is good for our atmosphere and humans do not need to reduce our 'Carbon Footprint' to save the earth or reduce pollution. In other words, the earth has warmed in the past, but this warming was caused without humans putting C02 into the atmosphere. Climate Change so far has been shown to be driven by natural causes. This group exists to promote these ideas and to 'cool down the freak out' with the Global Warming Hype." (Global Warming??, 2010)	Active

Source. Facebook.

Other examples of climate change groups on Facebook are the Arab Climate Change Group (88 members), the Unofficial Climate Change Deniers Group (25 members), Stop Climate Change (700 members), and Climate Change Sexy Action Group (1 member).

Facebook groups on climate change issues generally make extensive use of visual communication in the form of pictures and video clips. Among those most frequently used are images of industrial emissions and images of polar bears. These images are also widely used in the news media and can even be defined as icons representing climate change issues (O'Neill & Hulme, 2009). Some groups also use graphics from the scientific sphere—for instance, temperature models—a strategy that could be seen as a way to increase the credibility of the message. When it comes to the sources of information used, the analysis of the 50 groups shows that information comes mainly from members and is often in the form of links to websites—for example, to universities, scientific journals, and sites of scientific communities. In this way the information is framed as official knowledge. Links to video clips from electronic media (like television news clips) are also common communication tools in these groups. All in all, most of the information provided in these groups comes from rather traditional and official sources.

The following excerpts exemplify how the discourse of these groups is structured and how climate change is framed. They come from one of the most active groups, Join the Fight to Stop Global Warming, and more precisely from discussion on the topic "The Myths and Omissions of Global Warming," which included 586 conversation items (September 2010).[1]

Participant 1:

> Hi —: Good to see that you are still active in this board. My opinion hasn't changed too much. I am still living in a house that is only about twenty feet above sea level, and the road that I travel on to work is only about ten feet or less above sea level. I am really really really worried about rising ocean levels over the next few

decades. Frankly, I think that this guy Graham Hancock even has a point that I should be worried about a polar magnetic shift initiated by the massive recent buildup of ice in Antarctica!?
My opinion ... is basically that the central super cold region of Antarctica due to its massive size and altitude, has been acting a bit like a black hole. A general warming trend worldwide has increased levels of atmospheric H2O but as soon as a cloud goes over Antarctica, the clouds drop their load and as Mr. Gore mentioned in his film, the ice is then relatively stable for 650,000 years or more!

Participant 2:

—, Don't worry about Antarctica gaining too much mass. Though it may be true that higher altitudes, are getting increased precipitation, this increase is more than offset by increased melting at lower altitudes. Antarctica as a whole, has lost ice over the last ten years, at an accelerating rate. "In Antarctica the mass loss increased from 104 Gt/yr in 2002–2006 to 246 Gt/yr in 2006–2009, i.e., an acceleration of −26 ± 14 Gt/yr2 in 2002–2009."
http://www.agu.org/pubs/crossref/2009/2009GL040222.shtml

Participant 3:

20 ft above sea level is nothing to worry about! The reversal of the Earth's Magnetic Poles is expected to occur somewhere between 2012 and a million years from now according to scientists.... and if it does, so what? Like everything else, including global warming and cooling, there is nothing we mere humans can do about it!

To get a closer look at different discourses, I have conducted a comparison of how climate change was framed in one climate group (Join the Fight to Stop Global Warming) and in one anticlimate group (Global Warming??). The group Join the Fight to Stop Global Warming framed the main cause of the problem as consumption and lifestyle. Climate groups often focus on anthropogenic activities as part of the problem, and therefore a change in human behavior is part of the solution to climate change. Join the Fight to Stop Global Warming put forward use of electricity and cars as problem causes, and suggested ways that individuals could contribute with a lifestyle change. Typical of climate groups was the lack of uncertainty regarding claims: The assumption was that climate change is happening and action is needed now. The anti–climate change groups, on the other hand, questioned whether there really is a change in the climate and whether it really can be linked to human activities. According to the group Global Warming??, global warming should not to be seen as a threat to the planet and is not caused by human behavior. In contrast to the climate groups, Global Warming?? gave space to some uncertainty in that its members, to some extent, claimed to be open to the possibility that future knowledge could force them to change their opinions. The use of the time perspective in this way represents an uncertainty rhetoric that is also often used by scientists and journalists (Hornmoen, 2009).

I now return to the questions whether and how these media can be understood from a public sphere perspective. One of the main ingredients in the public sphere is interaction among citizens, civil society, and the state (Castells, 2008). In this sense social media such as Facebook cannot be seen as public spheres because much of the communication does not reach or involve the state. However, if a public sphere is defined as an arena for public discourse on current issues, these media definitely seem to have at least public-sphere potential. This exploratory analysis shows that many members in Facebook groups, for example, participate actively in discussions on climate change issues and use and share their professional and personal experiences in the discussions. The discourse in the groups does not necessarily involve the same actors as the mainstream (political) public sphere. In this sense, climate change groups on Facebook can be said to function as an alternative public sphere and perhaps also a green public sphere (cf. Yang & Calhoun, 2007). At the same time, it must be questioned whether these rather small communities with limited reach are "public" enough to be a public sphere.

Climate Governance and Virtual Public Spheres   177

Facebook can also be seen as an alternative arena for organizations, including environmental NGOs. Greenpeace, for example, is very active on Facebook. Founded in 1971, Greenpeace is a global civil society organization that addresses global (environmental) problems. It intensively uses media and other communication networks for its purposes, making it an illustrative example of the relationship between global governance, global civil society, and communication networks (Castells, 2008). As a global NGO, Greenpeace is mainly engaged in the production and diffusion of different symbolic forms related to the environment (Tsoukas, 1999). Greenpeace consists of Greenpeace International, based in Amsterdam, the Netherlands, and 28 quite autonomous regional and national offices around the world, with physical or virtual presence in over 40 countries in Europe, the Americas, Asia, Africa, and the Pacific (Tsoukas, 1999). According to the organization's website, it had 2.9 million paying members in January 2007.

Climate change is certainly among the main topics of interest for Greenpeace. On the home webpage of Greenpeace International, climate change is at the top of the list under the headline "Issues we work on." According to the information provided on Greenpeace International's website, climate change is definitely real, and there is no uncertainty concerning the importance of this issue; solutions are also provided without any doubts. "Everyone knows that climate change is real and happening now. What many people don't know is that we have the solutions in our hands" (Greenpeace International, 2010a). What is considered the problem in framing climate change is closely related to solutions, and the webpage provides a list of problems and one of solutions, each containing three aspects. The problems addressed are climate impacts, coal and rainforest destruction; under solutions, Greenpeace International chose to list climate science, energy (r)evolution, and the project Cool IT Challenge, which is about urging the IT industry to become a leader in climate activism and to provide the technological solutions needed to fight climate change (Greenpeace International, 2010b).

Greenpeace strives for visibility (of environmental issues) and in its campaigns works actively with different media platforms. The organization has been very successful in drawing attention in the traditional news media (especially for its campaign activities) and is actively committed to an online presence, mainly using social media like Facebook, Twitter, Flickr, and YouTube. Greenpeace's public campaigns often seek to reach popular culture through a media presence by making environmental issues culturally meaningful (Doyle, 2007).

How then does Greenpeace use Facebook? From the webpage of Greenpeace International there is a link to a Facebook page bearing the same name, which has been "liked" by 625,520 people (September 25, 2010). As on Greenpeace International's webpage, on its Facebook site the organization presents a list of issues on which it campaigns, and again, climate change is at the top of the list ("Stop Climate Change"). The site also provides a range of photos and links to YouTube and other websites. It has a log where Facebook users can participate in debate and discussion on climate change issues and other environmental matters. Those discussions are initiated and structured by the input of Greenpeace International. However, there are also 270 discussion threads, and these are initiated by individual members. Some of these concern the issue of climate change and also include comments from skeptics, as in this example from September 2010:

> We campaign on a variety of issues, centering on the following: 1) Stop climate change... climate is changing constantly ... we life [sic] on earth, not in big greenhouse!! the earth travels through space... you know, cosmos ... you cant even imagine what could drive the climate ... so instead of stop climate change, stop genetically modified species witch [sic] can do serious damage to our biosphere, or depleted uranium pollution, or pharmacological contamination ... green movement has been taken by communists ... sooo you should change you colors ... to red ... peace!
> (Facebook user, 2010)

Because of the great diversity and number of people participating, there is also a great diversity in how climate change is framed. Besides Greenpeace International's Facebook page, there are over 500 groups with some relation to Greenpeace, and of those 50–100 are using *Greenpeace* as some part of the group name and seem more or less official. For example, Greenpeace Sweden has 599 members, and Greenpeace has 125,065 members. The Greenpeace groups on Facebook are just as diverse as the climate change groups and vary widely in terms of number of members, level of activity, sources of information, and so on. There are not, however, any climate skeptic groups (although skeptics participate in the discussions) using the Greenpeace name. On most Facebook pages that relate to Greenpeace, the issue of climate change is acknowledged as one of today's most pressing problems.

As part of global civil society, Greenpeace, along with other (environmental) NGOs, is an important participant in the public sphere on climate change and other environmental issues, or what can also be called the "green public sphere" (Yang & Calhoun, 2007). NGOs play an important part in producing this critical discourse in which mass media—especially the Internet and alternative media (e.g., material from NGOs) —are the main means of communication. Facebook groups are of course only a small part of this communication but seem to function as an alternative platform for discussions on climate change, allowing the public (members of the groups) to participate in the production and consumption of discourse, which is focal in the green public sphere (cf. Yang & Calhoun, 2007).

Second Life: Environmental discourse in a virtual world
The phrase *virtual worlds* is generally used to describe an online software platform, providing its users access to 3-D spaces that function either as spaces for play or as extensions of "real life." Virtual worlds peaked in the beginning of the 21st century. Second Life, for example, was launched in 2003 and grew rapidly in terms of users, garnering much attention in public debate as many authorities and politicians attempted to be seen

as forerunners and to avoid being left behind. Lately, however, virtual worlds have not been receiving the same amount of attention, and the number of users is no longer increasing. Users are often very active and engaged but are not numerically impressive. Nevertheless, these virtual worlds are interesting examples of how the Internet and online media can be used, and they are here analyzed as such.

Second Life was developed and published by the American company Linden Lab and can be characterized as a user-defined world or space where users ("residents") interact and communicate through motional "avatars," or digital identities used in virtual worlds. When it comes to number of users it is difficult to find up-to-date information, but according to a blog post on the Second Life website, the number of users was 536,000 in 2007 and 769,000 in 2009. The same blog post states that the Second Life economy in 2009 totaled $567 million U.S. dollars (Linden, 2010). Previous studies of Second Life have analyzed, for example, health campaigns, issues of child pornography, copyright issues, crime and law in virtual worlds, consumption, public spheres, and business (e.g., Adrian, 2009; Huang, Kamel Boulos, & Dellavalle, 2008; Lin, 2007).

In Second Life (according to the presentation on the website www.secondlife.com), users can explore, meet other residents, socialize, participate in individual and group activities, create and trade virtual property and services with one another, and travel. A number of organizations, authorities, corporations, universities, media, and others have established themselves in Second Life. In June 2007, for example, the Swedish politician and then foreign minister Carl Bildt inaugurated the Swedish embassy in Second Life (Danielsson, 2007). The same year, the American news channel CNN launched an office in Second Life—an office that uses avatars (citizens) to do the job and create a form of public journalism ("CNN Enters the Virtual World," 2007). There is no fee to join Second Life, but money is required to participate in online transactions. In fact, virtual consumption has proved a very important component in Second Life usage. Second Life has also been used as an alternative arena for marketing real-life products; Toyota, for example, made virtual models of its new Scion car that users could test-drive in Second Life (Lin, 2007).

The exploratory analysis of Second Life presented here aims to identify elements that may contribute to understanding how this platform can be used as a space for communicating and deliberating on environmental issues. I conducted this examination in three steps. First, I carried out a general mapping of Second Life and its different parts and activities in order to identify to what extent and in what forms there were elements with relevance for environmental communication. This step was conducted through participatory observation and text analysis. Elements considered are places and actors with special relevance for the environmental and sustainability area, together with values and arguments related to sustainability issues as expressed in the presentation and description of different parts of Second Life. In the next step, the analysis focuses on the spaces in Second Life with the most references to the sustainability discourse, and on the questions (1) how environmental issues, such as climate change, are framed in those places and (2) how sustainability values are argued for. Frames are identified through a reading of certain communities' self-presentation (of, e.g., a place, an organization, an activity). The third part of the study focuses on Greenpeace and its activities in Second Life.

The first broad mapping of the Second Life structure and activities shows that although economic and cultural frames are dominant, there are also many references to the environment and sustainable development; such values permeate at least certain parts of Second Life. Better Earth, for example, is a space in Second Life that according to its own presentation wishes to "highlight new ways to improve our relationship with nature." It tells visitors to "explore the region for information on new agricultural innovations as well as sustainable food sources" (Better Earth in Second Life, 2010). Another example is the Etopia Eco Village, which is presented as an environmental sustainable living community focusing on issues like alternative energy and transportation, organic farming, recycling, global warming, and climate change. In the village there is a bike shop occasionally providing free bamboo bikes, and the Sustainable Living Library. The library is sponsored by the Colorado Association of Libraries (CAL) and works through workshops, exhibits, and other means to educate visitors about "green practices and other ways to sustain healthy lives and a healthy planet" (Etopia Eco Village in Second Life, 2010).

There are several examples of how virtual worlds like Second Life can be and are used as platforms and spaces for environmental communication. Organizations, as well as individuals, arrange and participate in events, discussions, consultancy, and other activities. One of the main communicative functions of Second Life so far has been to provide an arena for virtual meetings for people. Here Second Life merely works as a communicative tool. In 2007 about 5,000 employees of IBM, for example, were using Second Life, and this was among other things said to have replaced video conferences (Wallström, 2007). This function of Second Life is sometimes described as environmentally friendly, for it reduces the energy required to transport people to and from meetings (cf. Huang, Kamel Boulos, & Dellavalle, 2008). In this sense, transmission is said to replace transportation, and meetings become a form of virtual copresence. The OneClimate Island is a place in Second Life with links to the network OneClimate—which, among other things, offers to arrange virtual meetings. "If you are frustrated by not being able to fly, since flights are so carbon-heavy, here's a way to 'meet' people without flying. You can meet them on OneClimate Island, our virtual meeting space in Second Life" (OneClimate, 2010). Meetings and conferences can also occur simultaneously in real life (IRL) and in virtual worlds, and in these cases participants in both worlds are able to see each other and communicate and interact. The United Nations Conference on Climate Change that took place in Bali in December 2007 is one example of this (Huang, Kamel Boulos, & Dellavalle, 2008). "Hoping to widen the debate and cut down on carbon emissions from air travel, Oneworld.net, a left-leaning website, has taken the UN conference on climate change in Bali to the online virtual world Second Life" ("Second Life Climate Conference," 2007). Because it created opportunities for more people to participate (albeit virtually) in the conference and take part in the discussions, this event was certainly framed in public-sphere terminology as democratizing and a tool for deliberation: "Through Second Life, it doesn't matter if you're a farmer in Turkmenistan or a professor at Harvard. Everyone can ask questions anonymously and everyone's opinion is just as valid" (Allen, 2007, n.p.).

Especially interesting, perhaps, in the analysis of Second Life as a space and tool for environmental communication are the Commonwealth Islands, presented as a progressive and activist community formed by organizations and individuals using Second Life as a platform for promoting social and environmental changes.

> At ground level, the Islands are carefully crafted open space ecosystems that encourage community through exploration and shared experience. Meeting spaces (tree houses, sea-side amphitheaters, campgrounds, etc.) that are ideal for public and private events of all sizes are available to community members.
> (Commonwealth Islands in Second Life, 2010)

The Commonwealth Islands are also often related to a sustainability discourse, addressing both social and environmental as well as economic sustainability. "On Commonwealth, people from non-profit organizations build community and further their goals of positive social change, environmental sustainability, social and economic justice, peace and harmony with the earth" ("Greenpeace's Second Life," 2008). These descriptions of the Commonwealth Islands clearly show at least an ambition to function as a public sphere, allowing users and residents to participate in discussions and dialogues about environmental issues. Interestingly enough, there are indications of a return to the traditional Habermasian bourgeois public sphere in which citizens actually meet (albeit, again, virtually) to deliberate on public matters.

Figure 2. HippyJimStarbrook and Greenpeace in Second Life.

Source. Sunshine Communications (2008). Copyright 2011, Alan Rycroft (photographer), www.SunshineCommunications.ca, 250-592-8307 Canada. Reproduced with the photographer's permission.
Note. "March 29, 2008—Supporters of Greenpeace, including two UK staffers, meet near the organization's Second Life Commonwealth hut to discuss how to advance Greenpeace causes in virtual reality. Convener HippyJimStarbrook sits on the far left (long, black hair); reporter Alan Innis is wearing an Obama t-shirt" ("Greenpeace's Second Life," 2008).

To better understand how virtual worlds can be used as platforms for environmental communication and civic engagement, I now turn to civil society and Greenpeace's presence and activities in Second Life. Greenpeace is active in different parts of Second Life, including the Commonwealth Islands, where the organization has an information hut. Greenpeace also has a store in Ecotopia Prime. HippyJimStarbrook (UK) is the leader of Greenpeace in Second Life (figure 2). Users can meet with this avatar and discuss different issues.

Figure 3. Greenpeace's Dove campaign in Second Life.

Source. Sunshine Communications (2008). Copyright 2011, Alan Rycroft (photographer), www.SunshineCommunications.ca, 250-592-8307 Canada. Reproduced with the photographer's permission.

Besides providing opportunities for users and residents to approach them, ask questions, and discuss environmental problems, Greenpeace has also organized several events, such as Earth Day in 2007 and a virtual whale-hunt event in which people could ride a Greenpeace inflatable boat and participate in blocking harpoons. Greenpeace has also used Second Life as an extension of its regular campaigns and activities to broaden the arena for its communication. In relation to a campaign against Dove soap products, an opportunity for supporters to have their picture taken in Second Life with a statement against the manufacturer was made available (figure 3). When asked if this had any effect IRL, representatives for the organization said that it "added to the 'numbers pressure'" ("Greenpeace's Second Life," 2008).

Thus Greenpeace has clearly recognized that virtual worlds like Second Life can be used as platforms for communication and deliberation. The general pattern is that large organizations like global NGOs tend to think of virtual worlds as a tool to reach an audience without the constraints of national boundaries (Adrian, 2009), but the organizations' main efforts are not invested in these media given their rather limited reach and usage. For now, Greenpeace's online communication strategy is mainly about using social media (e.g., Twitter, YouTube, Facebook).

Climate change is not a main part of the agenda in Second Life, even if there are references to the issue in the context of places and events related to environmental or sustainability discourse. The Etopia Eco Village, for example, presents climate change as an issue that is recognized and addressed in the community ("sustainable living for stopping climate change"; Etopia Eco Village in Second Life, 2010). In spite of these occasional climate change references, however, the general impression is that environmental discourse in Second Life is not so much related to specific issues but is rather an expression of more general values on the need for sustainable development.

What can one conclude on the role of Second Life as a communicative space for environmental discourse and deliberation? Because actors in Second Life and other virtual worlds consist of avatars and not "real" people, this platform is not really a suitable tool for a politician who wishes to increase his or her visibility in the public arena. In this respect, Second Life is perhaps a better tool for making organizations visible. However, political campaigning does occur, and different discussion forums are available. It is also obvious that virtual worlds like Second Life can function as alternative meeting spaces for deliberation on different matters. Thus Second Life clearly has some public-sphere characteristics. "Every new medium reflects and represents the world in a wholly new way.... Second Life ... is a surreal, idealized environment created

Climate Governance and Virtual Public Spheres 187

by and inhabited by real people having real conversations about real issues" (Rycroft, 2008). Thus Second Life can work as a virtual meeting space and a virtual sphere were citizens/avatars meet with others and engage in discussions without having to take issues of distance, power, status, and so forth into consideration. This in many ways moves away from the idea that today's public sphere is mediated (in the traditional sense—that is, involving journalists and other communication professionals); the development bears some resemblance to the Habermasian ideal bourgeois public sphere.

ONLINE MEDIA AND ENVIRONMENTAL DISCOURSE: REMAKING THE POLITICAL?

This chapter has examined ways that environmental risk governance and politics might be transformed with new forms of public participation and communication associated with the emergence of global communication networks and potential online public spheres. To sketch societal transformations up to the current network society, the world has seen a transformation of public-sphere arenas from the coffeehouses in Habermas's bourgeois public sphere to modern mass media and journalism, and to mediated virtual coffeehouses characterized by a virtual copresence. Virtual public sphere(s) can in many ways be said to work as imaginary public spheres; their main function is not to gather citizens in rational discussions and deliberation on public matters but rather to create a sense of community and engagement. Just like the ideal Habermasian public sphere, these virtual public spheres are inclusive and accessible (for users/citizens) and allow for a practice of citizenship not linked to national boundaries. According to Nancy Fraser (1992), a public sphere can be "alternative" in various ways. It can be different in terms of where the discourse takes place (in an alternative arena rather than the mainstream mediated public sphere); in terms of who gets to participate in discourse (not those who normally dominate the mediated public sphere); in terms of the issues being addressed (issues other than those commonly debated in the mainstream public sphere); and finally, in terms of what forms of discourse and debate are used (for example, forms encouraging citizen participation). Seen this way, both social media like Facebook and virtual worlds like Second Life exhibit several characteristics of an alternative public sphere: the discourse takes place in a different arena (online) and is more open and inclusive (especially for the young and digitally equipped). However, in terms of the issues addressed and the framing of issues like climate change, these platforms work rather like extensions of the mainstream mediated public sphere.

To fully understand the role of online media for contemporary politics and public debate, it might also be fruitful to distinguish between what can be called the *political public sphere* and the *cultural public sphere*. Habermas was mainly interested in the public sphere as a locus for political power, whereas Nancy Fraser, for example, discussed the role of the public sphere as an arbiter of cultural recognition (cf. Fraser, 1992). Discussions of the role of news media and their importance in society and for democracy are often related to the former kind of public sphere. However, some researchers claim that online media and user-generated content generally have a democratic potential in terms not of political influence but of cultural equality (cf. Jönsson & Örnebring, 2011).

Because the focus here is not democracy and public discourse in general but environmental governance and current problems like climate change, it is also of interest how these global communication platforms can be understood in terms of a "green public sphere" (cf. Yang & Calhoun, 2007). Obviously, social media and virtual worlds are not entirely devoted to discourse on environmental issues or "greenspeak" as Yang and Calhoun (2007) defined it. Still, owing to their global character, online social media and virtual worlds seem to work well for deliberation and discussion on environmental issues like climate change. Another relevant question is how these media can be understood in the context of the discussion of sustainable communication. On the one hand, it is claimed that virtual worlds and online spaces can serve as alternative meeting places and thus reduce emissions from air traffic and so on. Modern consumption has often been described as a cause of climate change, and virtual worlds can be said to offer alternatives to real life and unsustainable consumption (Lin, 2007). Thus the potential contribution of online media toward a sustainable society has to do not only with how they can be used as tools for communication and deliberation but also with helping people change behaviors that affect the environment negatively (e.g., reducing consumption). In contrast, server empires like Google, YouTube, Second Life, and Facebook can also be seen as a threat to the environment because of their extensive use of electricity (Rittsel, 2007).

To sum up, online communities do offer alternative venues and spaces for environmental communication, are more inclusive and accessible than mainstream media, and give some expression to the idea of a global citizenship. In this respect, these media can be seen as alternative public spheres and part of a green public sphere. However, these arenas are open to (almost) everyone but not used by everyone. They have failed to attract a representative sample of the population (cf. Dahlberg, 2001b). So the answer to Zavestoski, Shulman, & Schlosberg's (2006, p. 2) question—whether the Web can provide an "improved arena for democratic deliberation" and an online public sphere—is yes. But it does not yet live up to its full democratic potential. This in many ways aligns with Dahlberg's (2001b) conclusion that in order to function as an expansion of the public sphere, the Internet needs to not only provide deliberative spaces but also attract citizens (especially the young) and encourage participation from all groups in the population (Dahlberg, 2001b). Dahlgren (2006) argued that civil society may serve as a training ground for civic participation, and the same could be argued for online virtual media. Activities in these media are also a way of gaining what Dahlgren called "communicative competences" and what could also be described as "civic literacy." Society is perhaps approaching a time for the "citizen avatar." It is also worth noting that online media work as virtual public spheres mainly for young adults and youths. Through the use of these media, young people are building a new political culture (see, e.g., Rycroft, 2011). If this is considered a generation issue, the importance of online media and virtual public spheres will probably increase with time and further change the terms for deliberation and communication on environmental issues.

Dahlgren presented an argument for a "cultural turn" in studies of citizenship and civic agency. The idea of deliberation should be widened in terms of process as well as of content in order to better suit the contemporary media milieu and political context. As for process, deliberation should include not only rational communication, talk, and dialogue but also interaction in itself; and in terms of content, the kind of talk that should count as deliberation should include everyday, informal talk and discussions with links to the private sphere (cf. Dahlgren, 2006). According to Dahlgren (2006), in the field of politics public spaces are intertwined with private spaces, and this becomes very clear if one looks at social media and virtual worlds.

According to Castells (2008), the growing gap between the space where problems and issues arise (global arena) and where they are handled (national arena) has resulted in four political crises that affect governance institutions: a crisis of efficiency, a crisis of legitimacy, a crisis of identity, and a crisis of equity. Virtual worlds and social media are perhaps not today proper governance instruments, but they can certainly contribute to the emergence of a public sphere that is not related to the national arena alone and is therefore better suited for late-modern governance of a sustainable society, providing democratic potential in terms not of political influence but of cultural equality. There seems to be a need for a sustainable public sphere in which sustainable means of communication are used to address and deliberate upon issues of ecological and social, as well as economic, sustainability.

Endnotes

1. I have omitted names and images of participants.

CHAPTER 7

WEB-BASED PUBLIC PARTICIPATION

A CASE STUDY OF 350.ORG

Andrea M. Feldpausch-Parker,
Israel D. Parker, and Tarla Rai Peterson

INTRODUCTION

The topic of anthropogenic climate change has become a quagmire for both national and international politics. When it comes to the creation of climate policy, few if any nations can boast a comprehensive climate plan that has not run into significant opposition from established interest groups (Aldy, Barrett, & Stavins, 2003; Bohringer, Loschel, Moslener, & Rutherford, 2009; Stavins, 2008). Top contributors to anthropogenic climate change, such as the United States and China, have avoided setting climate policies because of the potential price tag, political ideology, and associated uncertainty (Wang & Watson, 2008). To circumvent some of the political stalemate in the United States, community leaders and policy makers have attempted to use federal-level mechanisms already in place, such as the Clean Air Act and fuel-efficiency standards, and to create subnational and municipality-level climate policies to move forward in reducing greenhouse gas emissions (Goulder & Stavins, 2011; Sharp, Daley, & Lynch, 2011). Countries have also stalled comprehensive climate change legislation because of international disputes over relative responsibility for anthropogenic climate change and the costs of mitigation and adaptation (e.g., developed versus developing nations; Bosetti, Carraro, & Tavoni, 2008; McKibbin & Wilcoxen, 2002). Some nations, such as members of the European Union, however, have been relatively successful in their national climate policy efforts and have already set targets for reducing their greenhouse gas emissions—a success considered achievable because of networked governance between the nation-states (Giddens, 2009). What has been overlooked, however, by many policy makers is the support from citizenry and how this support would provide the public with the necessary footing for creating effective policies.

As climate policy experts battle over how to mitigate and adapt to anthropogenic climate change, the public tends to be relegated to the sidelines (Feldpausch-Parker, 2010; Rayner, 2003). Although this technocratic approach is somewhat understandable because of the complexities of the issue and its heavy reliance on scientific information, it disregards the potential contributions of the lay public to climate policy. When citizens attempt to participate in public events related to anthropogenic climate change, they simultaneously conform to and change the rules for participation in policy debate. For the most part, governments have failed to harness the potential power of the public in influencing the decision-making process to address climate change (Hyde, 1990; Laird, 1990). Climate change campaigns like those orchestrated by 350.org, however, demonstrate global citizen support for further international treaties and joint policy making, and have inserted the voice of the public in climate negotiations.

In this chapter, we examine how organizers for 350.org attempted to provide opportunities for public participation in institutional mechanisms intended to forge international climate change policy. Using event announcements as a source of data, our analysis focuses specifically on how event organizers utilized (1) material and human resources and (2) institutional rules and policies in order to assert authority and put forth the 350.org message of action to global policy makers. We first provide background on the emergence, goals, and strategies of 350.org; this is followed by an outline of our theoretical framework for analyzing event announcements. Next we present summaries of event announcements, and finally discuss our findings regarding the implementation of the resources/rules rubric as a means for judging how the events' communicative and policy potentials were realized.

THE HISTORY OF 350.ORG

On April 14, 2007, Bill McKibben and a group of students from Middlebury College used the Internet to successfully assist in the online organization of 1,400 community-based events in all 50 U.S. states with a single, united message—"Step it up Congress, we want 80% reductions by 2050," a rallying cry meant to unify the otherwise disparate events (Endres, Sprain, & Peterson, 2009). McKibben and his students used the Web as a communication medium for organizing a mass do-it-yourself campaign to encourage multiple citizen protests that enabled local event organizers and other participants throughout the United States to envision possibilities for local influence on national climate policy (Endres, Sprain, & Peterson, 2009). They managed to capture the attention of numerous politicians across the United States, including participation by both state and national senators and representatives (and presidential hopefuls), when the events were repeated roughly six months later on November 3, 2007 (Step It Up, 2007).

In preparation for the 2009 United Nations (UN) Climate Change Conference in Copenhagen (December 7–18, 2009), the Step It Up organizers decided to go global with a refined message—reduce atmospheric

CO_2 concentrations below 350 parts per million—a number suggested by many climate scientists as the safety limit for humanity's continued thriving on Earth (350.org, 2011). Transitioning from the national to the international arena with a more technically precise and easily translatable message, McKibben and his newly dubbed 350.org team again used the Internet as a tool for mass online organization.

As a nongovernmental organization (NGO), 350.org provided participants with resources for developing knowledge on the topics of climate change science and policy. One of the most obvious ways 350.org organizers accomplished this goal was by using the organization's website as an educational tool, providing access to information regarding (1) climate science, (2) rules of public engagement, and (3) policy options. The 350.org website served and continues to serve as a hub for information on multiple actions geared toward affecting policies related to anthropogenic climate change. This includes providing basic information about the organization, their mission and various news feeds, as well as an online instruction manual for climate campaigns, organizational archives, and a donation box. The instruction manual not only provides Web space for an event, complete with template, but it also provides information to local event organizers on how to communicate with the media (e.g., "Learn the 350.org and Global Work Party Talking Points," "The 3 Steps to Great Media Coverage for Your Global Work Party Event," in 350.org, 2009j), blog, take photos (for later posting on the website), make a video, and create a report for an event. The 350.org team members went beyond the website, promoting their message through multiple forms of social media, such as YouTube, Facebook, and Twitter, thus expanding their reach into cyberspace and possibly increasing the number of their cells by going "viral." Access to this knowledge, therefore, allowed local event organizers and participants to develop a degree of expertise on climate change campaigning and thus provided a pathway for inserting the public into public-policy discussions.

On October 24, 2009, roughly two months before the Copenhagen summit, the 350.org organizers worked with local event planners to orchestrate the production of 5,245 separate events in a total of 181 countries around the world (350.org, 2011). As events unfolded, the NGO had local organizers post images from the events on the 350.org website as an additional demonstration space in order to deliver the message to national and international policy makers. The goal of these events was to "unite the world around solutions to the climate crisis" (350.org, 2011). This global voice represented diverse peoples, ecosystems, cultures, and governments in an attempt to promote global cooperation by all nations in climate action in Copenhagen. By demonstrating unified citizen support around the globe, 350.org organizers and participants hoped to transform government inaction and passivity into a flurry of policy activity at the Copenhagen summit. Though many argue that the summit fell short of any type of cooperation by participating countries (Pendleton, 2010; Rogelj et al., 2010), the 2009: International Day of Climate Action proved that there was no shortage of support from a global citizenry.

In this chapter we use Giddens's (1984) structuration theory to examine how an Internet-based campaign facilitated the insertion of the public into international climate policy making. We focus on a set of texts and images posted on the 350.org website from the countries of Austria, Madagascar, the United States, Venezuela, and Yemen in order to examine how people who are based in countries with diverse governmental structures and who face a broad range of hazards associated with anthropogenic climate change used the Internet to insert their concerns into climate change policy. We also highlight how Web-based communication has changed the way human beings communicate environmental issues and create social movements. This organization provides a useful example of how members of the lay public may use the Internet as a communication medium to make a statement about global biophysical processes in a manner that is adapted to local political realities.

STRUCTURATION THEORY AND PUBLIC AUTHORITY

Giddens's (1984) theory of structuration represents an attempt to provide social scientists with a theoretical framework for analyzing society that simultaneously accounts for both structure and action, especially as articulated by social system and individual agent. He posited that society is dependent upon both the individual agent and the collective; without one, the other cannot exist. Individuals, as agents, are responsible for the continued reproduction of system properties that drive society. This idea presents a duality of structure, in which "the structural properties of social systems are both medium and outcome of the practices they recursively organize" (p. 25). This reproduction of social systems that make up social structure both constrains and enables individual agency. Even though humanity functions within the confines of the social systems human beings themselves have created, it does not necessarily prevent people from addressing the need for change. Realizing the potential for change, however, must still be achieved within the social systems already in place.

Applying this theoretical framework to citizen involvement in efforts to influence climate change policy and action, Giddens described how citizens may influence political processes through their utilization of *resources*, or "media through which power is exercised, as a routine element of the instantiation of conduct in social reproduction" (p. 16), and *rules*, or "generalizable procedures applied in the enactment/reproduction of social practices" (p. 21). Thus resources and rules serve as the essential components of structure and power. Norton (2007) noted that structuration theory is founded on the proposition "that social relations are constituted by people (agents) with *agency*, or the ability to choose otherwise" (p. 150) As Norton suggested, structuration provides guidance for an analysis that accounts for both systemic issues and their manifestation through micropractices. This is particularly useful when examining campaigns such as 350.org, which operate at intersections between globalization and localization, between symbolic and material forms, and between organizational control and individual resistance.

In the case of 350.org, new media (e.g., websites, social networking sites, YouTube videos) enabled a vast collective of people from around the globe to assert a type of *public authority*, or representation of the public's needs and reasonable demands upon their leaders. Any structure's resources and rules enable a dialectic of control wherein both individual agents and social institutions exercise varying levels and forms of power (Norton, 2007). Citizens of spatially disparate locales can influence the actions of other individuals and institutions. The organizers of 350.org, then, work within existing structures, such as legal protections afforded to NGOs and the growing use on the Internet as a means of operating, in order to encourage individual agents to engage in concerted action intended to change those very structures. They offer guidance for using existing resources and rules to shift power away from traditionally influential gatekeepers to previously absent or silent participants in the climate change debate.

You're invited to an event!

In this section we highlight descriptions posted on the 350.org website by participants in 17 different events in the countries of Austria, Madagascar, the United States, Venezuela, and Yemen. We are primarily interested in examining how event organizers chose to present their events so as to encourage participation and in audience recognition of the 350.org message through the use of rules and resources as a means to assert public authority over the climate change negotiations in Copenhagen (table 5). We have organized our analysis to determine how 350.org attempted to achieve greater public authority over climate change issues using a framework formed from Giddens's (1984) structuration theory. We selected the countries included in this study to showcase the wide breadth of government structures, continents, and direct exposure to immediate impacts from climate change. We drew the 17 announcements for analysis, along with accompanying images, from groups that formed in the five countries. All announcements were posted on the 350.org website prior to the day of the proposed global event (October 24, 2009).

Event organizers responded to the global challenge of anthropogenic climate change by using regionally and culturally based strategies in order to provide a voice to their national citizenry. To adequately evaluate these unique methods, we relied upon seven descriptive characteristics that illuminated different governments, knowledge systems, cultures, and unique regional or national vulnerabilities to anthropogenic climate change. These seven characteristics were intended to provide insight into how local organizers attempted to use regionally or nationally specific social structures (rules and resources) to encourage their national representatives to push for comprehensive climate policy in Copenhagen. Analytical categories used for all events were (1) geographic location, (2) government regime, (3) knowledge systems, (4) political or popular figureheads, (5) iconicity, (6) sponsorship and/or collaboration, and (7) audience and participant targeting strategies.

Austria

Event 1: Stephansplatz, Vienna

Vienna's Stephansplatz is deeply significant to Austrians; it serves as Vienna's center and is home to one of the world's tallest cathedrals. Although not expressed distinctly in the announcement, the choice of Stephansplatz reflected more than logistical ease. Economically, this square is a central location for shopping, dining, and tourist traffic. Politically and culturally, it has served as rallying point for Austrians to express opinions on a variety of national and international causes and as a location for festivals. In addition, Vienna is the capital of Austria;

the location therefore created a sense that participants had a direct line to the national government.

Table 5. Political demographics.

Country	Continent	Population (as of 2010)	Total Area (in sq km)	Coastline (in km)	Type of Government	Engagement with Copenhagen Accord
Austria	Europe	8,214,160	83,871	0	Federal Republic	Associated with target
Madagascar	Africa	21,281,844	587,041	4,828	Republic	Submitted actions
United States	North America	310,232,863	9,826,675	19,924	Constitution-based Federal Republic	Associated with target
Venezuela	South America	27,223,228	912,050	2,800	Federal Republic	Will not engage
Yemen	Asia	23,495,361	527,968	1,906	Republic	Yet to respond

Sources. U.S. Central Intelligence Agency (2011); USCAN Climate Action Network (2010).

The local organizers sought to pressure Austrian representatives to act on the international stage because their leaders were "negotiating the most crucial climate treaty in the history of humankind this December and their level of ambition is far below safe limits" (350.org, 2009l). Organizers expanded upon the international vision of unity, for the event became an invitation "to join thousands of people worldwide in expressing our concern for global warming." Interestingly, the Vienna event alone of the analyzed Austria events mentioned science in the climate change issue: "350 ppm [parts per million] is the concentration which is the safe limit for the survival of humanity according to scientists" (350.org, 2009l).

Included with the Stephansplatz event announcement was an image of a human banner and a sketch of a tree. Surrounded by buildings of commerce and government, a group of concerned Austrians brought the cement of the famous public square to life; roughly 60 people formed the outline of a person with arms outstretched, framing the number 350 (figure 4). In the enthusiasm of the actual gathering, the group demonstrated the unpredictability that accompanies activities undertaken by individual agents who can make choices and did not construct the originally intended tree, but a human form instead. Small patches of cement-enclosed green space framed the centered and symmetrical tracing of the stylized human figure, with additional people flanking the bottom. This representation highlighted the argument that human beings can and should accept responsibility for mitigating the effects of anthropogenic climate change. The perspective of the photographer captures the broader scene and also collects the entire group as one voice, one political directive for representatives to follow. The image provides a synopsis of the event for the audience that extends from the participants themselves to the capital and the government.

Event 2: Garden Party, Vienna
Although advertised with a relatively short announcement, the second Austrian event reflected a local and personal element in the form of a garden party. The goal, therefore, seemed to promote interaction among like-minded individuals in a comfortable setting. The absence of set speakers, national or international goals, scientific statements, and slogans reinforced the perception of a nonconfrontational approach.

Event 3: 350 Park Stroll, Salzberg
Like the garden party, the 350 park stroll seemed intended to internalize the message rather than share it with a broader audience.

Figure 4. Human banner from the Stephansplatz rally in Vienna, Austria.

Source. Image retrieved from the 350.org website, where it was posted by rally organizers.

Organizers of this event made no mention of communication outside their target participants, although a group walking silently through a public park would have been likely to communicate some sensibility to nonparticipants, even if the message lacked somewhat in clarity. Without specific signs or verbal communication, this strategy ran the risk of not making sense to those uninitiated into the event's meditative and contemplative approach. Organizers of this event sought to use 350.org's interest in climate change policy to demonstrate support for a universal concept of environmental awareness.

Event 4: Hauptplatz, Baden
The stated purpose of the Hauptplatz gathering was to "wake people up" and create a "change in climate politics." Beyond awakening people, however, goals or objectives were vague. The people to be awakened appeared to be the local citizenry most likely to hear the loud drums of protest, along with the Austrian politicians most directly involved in climate politics. The organizers emphasized the potential power of local citizens to effect change in policy by offering locally grown apples. This one-apple-at-a-time methodology highlights the role of individuals in reducing greenhouse gas emissions.

Madagascar

Event 1: Waking-up-Call on Climate Change, Antananarivo
Unlike many of the events mentioned in this chapter, the announcement for the Waking-up-Call on Climate Change event was fashioned less as a description of an individual event than as a more general statement of national outreach. The organizers partnered with a telecommunications company (TELMA) to spread their message about climate change. Part of this collaboration included calls to all TELMA customers alerting them about climate change in Madagascar. Organizers attracted speakers, including the minister of the environment and forests and the general director of TELMA/Madagascar, and had even secured an

announcement on broadcast TV. In keeping with its general outreach tone, the announcement contained few details traditionally associated with an event. Although urging a wake-up call on climate change, the website announced that the Waking-up-Call event would provide an opportunity to discuss "climate change and the future climate action in Madagascar" (350.org, 2009k). The goal was an informed citizenry of Madagascar united in its efforts to fight climate change instead of an explicit directive to support the goals of 350.org.

Event 2: Global Day of Climate Action, Antananarivo
With a stated goal to "reduce CO_2 emissions and get the PPM back to 350," the organizers of this event chose the pilot site of an environmental project. Although the choice to gather at a pilot community (near or in Antananarivo, the capital of Madagascar) likely had powerful local impact, the goal was national in reach, target audiences ranging from girls in rural areas (one co-organizing group was a member of the World Association of Girl Guides and Girl Scouts), government decision makers, and the Malagasy population as a whole. Thus organizers focused on power centers located in the capital of the country as they marched from a primary school to the queen's palace in the capital. In keeping with the nationalistic theme of the event post, the event was to include the singing of the national hymn and speeches by officials, followed by interaction with these officials. The organizers also concertedly worked toward involving all age groups and genders in drawing attention to climate change. They seemed to value not only direct communication with elected representatives but also inclusion regardless of standing. Finally, the event would attempt to relate the impacts of climate change on Madagascar in particular, provide information about climate change in general, and explain the significance of 350 parts per million. Throughout the announcement, the event was framed as a national day of unity supported by education and action.

Event 3: 350 Madagascar, Ft. Dauphin

This was an extremely short but powerful announcement relying upon those likely to be first affected by climate change to form the core of the movement. The organizers of the 350 Madagascar event wanted to "demonstrate the power and commitment" of what might be a large class of people in Madagascar. Although this was not specifically spelled out, those who most often feel the negative impacts of skewed environmental justice are the poor and the disenfranchised. That may well be the case in Madagascar, and a call for organization "of those who are, and will be, some of the first to feel the effects of climate change" may also reflect the mobilization of a specific class or classes in Madagascar (350.org, 2009c).

Event 4: Climate Change Resilience Building, Ranobe

The group New Latitude focused on *ho avy* ("the future") when pressing "long-term systemic solutions" to environmental problems in Madagascar. Though climate change is an international issue, this group focused on national and local issues in Madagascar, such as deforestation and the protection of the Spiny Forest, with the concomitant goal of promoting social and environmental justice. They sought to maintain the "inherent livelihood qualities of Malagasy people" by committing to "long-term sustainable community resource management" (350.org, 2009h). The focus was on continuity in the Malagasy way of life, but through a new, environmentally sound approach. A large component of the announcement informed the public about nationally important environmental issues and the steps that organizers had taken to combat them. The group demonstrated its hands-on and local approach by planting 350 seedlings to support 350.org. The organizers sought to intertwine the local needs and cultural identity with the larger issue of lowering the atmospheric concentration of greenhouse gases.

United States

Event 1: 350 from the Space Needle, Seattle

The Space Needle is a distinctly American building and an icon of Seattle, Washington. By meeting next to the Space Needle, organizers attempted to connect the urgent message of 350.org to a basic and visceral component of Seattle and the nation. This was reinforced by the participation of a variety of local advocacy groups. The organizers further connected the local to the national by promising to provide "3 things you can do in your home and community for greener living, for your children and for the planet" (350.org, 2009b), thus emphasizing the emotional consequences of inaction by implying that "your" children will pay the consequences. Organizers then linked the international events in real time by providing live streaming of other 350.org events. Finally, the overall announcement placed the onus on each participant through a variety of suggestions and statements including the following: (1) "come by human power," (2) "This is a pack-it-in pack-it-out zero waste event," and (3) "live feed of YOU taking action in your communities" (350.org, 2009b).

Event 2: GAIAliscious: Linking Cities to Climate Action, San Juan, Puerto Rico

During a multiday series of events, this event's organizers placed a great deal of emphasis on community involvement and storytelling (specifically about global warming). In an image accompanying the announcement, a woman was pictured holding a sign referring to her locality that simply stated, "*Quiero un Rio Piedras unido en su diversidad*" ("I want a Rock River united in its diversity"). Workshops about vegetable cultivation, ecomarkets, waterways, and artistic expression were organized in an effort to promote discussion and education about global warming in general and individual and community action in particular. Although mostly a schedule of events, the announcement placed the responsibility for change on individuals and communities.

Event 3: 350 Prayer Flags, Denver

This event proposed the amalgamation of spirituality, patriotism, and community with no explicit mention of the environment. Despite the patriotism, the commitment to community was paramount. As "churches, schools, and community groups [participated] in the gathering of prayers," organizers would display the collection of prayer flags (350.org, 2009d). Interestingly, the only mention of environmental issues was the use of the number 350 and a link to the 350 Denver website.

Event 4: Ann Arbor Fall Reskilling Festival, Ann Arbor

The premise for this gathering was the reeducation of the American workforce for new "green" jobs that will help transition unemployed or underemployed workers to long-term employment. The state of Michigan particularly suffered from lost manufacturing jobs, so the combination of reeducation with a new green focus made sense to the organizers (Transition Ann Arbor and Rudolf Steiner School of Ann Arbor). The announcement stated, "The Transition Towns movement seeks to build community resilience in response to the challenges of peak oil, climate change and the economic crisis" (350.org, 2009g), thus intertwining climate change with individual, local, and national issues (e.g., unemployment).

Venezuela

Event 1: Candle Light Event, Caracas

This candlelight vigil was among the shortest events, at 10 minutes long. Traditionally, candlelight vigils are used to support the bereaved or the imperiled or to draw attention to matters of social importance. In this case, the vigil was performed to highlight global warming while saving energy by turning off electric lights. This combined the immediacy and individuality of environmental awareness (e.g., a person turned off lights) with the concerns of a greater community (e.g., "Connect with thousands of people around the city who will be doing the same thing" (350.org, 2009a). The act of turning off the lights was supported by science: The announcement stated, "Twenty-one percent of the CO_2 produced by the planet comes from the electrical consumption of our homes, so this will be a big breath of relief that you will be giving the earth on day 350." The effort to conduct such an action in the capital of Venezuela might also have been a call to the representatives of the nation to support the goals of the organizers. Finally, an effort was made to humanize the earth: "Planet Earth and all its inhabitants will thank you." The organizers thus attempted to intertwine personal responsibility with global problems and supported the action with science and appeals to unity and morality.

Event 2: Fundación Código Aventura For Our Planet, Tumero
The organizers of this event connected the idea of peace with global climate change. They urged participants to "bring white flannel as a PEACE symbol" when they met for a "Conservatory on global warming and the climatic situation in Venezuela" (350.org, 2009i). With these statements they connected very important themes including nonviolence, global climate change, and specific climate change impacts on Venezuela. Although the number 350 was mentioned, it was not emphasized; organizers instead focused on the more general ideas of global warming or climate change.

In an image featuring 17 young people (eight girls and nine boys) with one adult, participants took a stand for future generations. The youth posed for the photographer against a black background, bearing flash-signs that might be part of popular culture or part of a group symbol. Nothing in the image indicates 350 parts per million, climate change, or the place except that local organizers posted the image on the 350.org website.

Events 3 and 4: 350 Venezuela – Committed to a Better Climate, Catia la Mar and Maracaibo

The VITALIS NGO collaborated with other NGOs (350 Venezuela, FEZU, and JGH University) to organize two events in different locations in Venezuela. The events had the same structure and message. They used the 350 number often, including in the organizers' goal of rallying 350 young students and citizens. The goal was to "centre the national, regional, and global attention about the [sic] climate change and the need of an urgent action" (350.org, 2009e). Organizers hoped to do this through a stark, scientifically backed summary of the impacts of rising sea levels on the Venezuelan shoreline. In an interesting activity, they planned to use "people's bodies [to] mark the new coast line as consequences of climate change." The groups hoped to garner national media attention for their events in order to further the 350.org message and to emphasize that Venezuelans were "citizens of the world" and obligated for both personal and global reasons to act to mitigate global climate change.

Yemen

Event 1: 350 Yemen, Sana'a

The event was held in the capital of Yemen, Sana'a. The organizers planned to march in the old city and to "highlight the importance of the negotiations and the role of Yemen in the process and the decision taking part in saving this planet or not" (350.org, 2009f).

They invited the minister of the environment to speak at the gathering, though it was unclear in the announcement whether the minister had committed to attending. The announcement, though short, demonstrated the organizers' endorsement of the goals of 350.org and the important role that Yemen could play in mitigation strategies.

Figure 5. Writing in the sand from the Yemen rally.

Source. Image retrieved from the 350.org website, where it was posted by rally organizers.

Using an image of the Yemen coastline, organizers focused on a fluctuating boundary of water and land, a major rhetorical appeal from the organizers of Yemen's only event. Shaped from the volcanic stones of the shore, which contrast with the lighter sands of the beach, the event's title is simply marked by the number *350* and by the political unit of efficacy, *Yemen* (figure 5). This simple composition overcomes the limitation of language, portraying a complex system—the ocean—and its link to humanity. Though not a politically overt message, it connected with the place's vulnerability and need for action.

IMPLEMENTATION OF RULES AND RESOURCES

Giddens's (1984) definition of rules highlights techniques and procedures as a means for continued enactment or reproduction of structure through social systems (i.e., political, legal, and economic institutions). In this chapter, we focus part of our evaluation of the use of rules in influencing the creation of national and international policies on climate change. By examining national characteristics, such as the type of government regime and the geographic location of a country, we found that few of the events sampled used strategies specific to their country's location (e.g., presence in the European Union or location on the Arabian Peninsula), government structure, or national standing within global politics. Many local organizers did, however, evoke patriotism in their announcements as an appeal to government authorities within legal institutions—for instance, asking Malagasy event-goers to sing the national anthem. Others made linkages to national needs and issues that held environmental and social consequences, such as a desire for green jobs (the Ann Arbor, Michigan, event) or disputes over class and environmental justice (the Madagascar events). Serving as an outlier, one Madagascar event seemed to be cosponsored by the Ministry of the Environment and Forests and the general director of TELMA/Madagascar, actors that essentially coopted the event so as to serve a more national agenda.

In addition to demonstrations of patriotism, some local organizers took advantage of their proximity to political centers as iconic locations in order to strengthen their message and better share it with their target audience, policy makers, in an appeal to political and legal institutions as the relevant social systems for acting upon climate change. A handful of the events were set on the steps of capital buildings or invited government officials as political figureheads to take part (Hauptplatz, Austria, and Global Day of Climate Action, Antananarivo, Madagascar). Other events focused on speaking *for* a community *to* government officials (San Juan, Puerto Rico; Caracas, Venezuela). Most of the events, with the exception of a few with more internal messaging (i.e., for participants or the local community), attempted to take advantage of the government's interest in current environmental issues in order to make connections with their national decision makers and demonstrate that climate change policies are overall advantageous to the country's well-being.

Overall, event organizers alluded to existing rules without being explicit in the rally announcements. This may have resulted from an agreed-upon understanding of the importance of these institutional structures among organizers and rally participants, or it may have resulted from a cursory understanding on the part of event organizers of how to use these institutions to their advantage in a campaign. Appeals to government institutions as a surrogate for social rules were common throughout the event announcements we reviewed.

The second component of structure, as Giddens (1984) explained, is the utilization of resources—both material (e.g., tools, money) and symbolic—as a means of shifting power structures. For the 350.org events, this included access to and use of event locations significant to the audience, invited speakers, sponsors, and (in some instances) other events. Some of these items overlapped with what we have constituted as rules. This is because institutions, which function under a set of laws and policies, are often a source of or have immediate access to resources such as knowledge, political ties, and funding (Norton, 2007). A majority of the events in our sample took advantage of iconic locations, sponsorship, or pairing with other events focused on the environment (e.g., green living). For example, Seattle's event took place at the base of the Space Needle, whereas Austria used public squares to gain voice. Two Venezuela events and the sole Yemen event used coastlines to portray the effects of anthropogenic climate change on their countries. Other groups used local, national, or international organizations and popular or political actors to further their messages—the Girl Scouts and the minister of the environment and forests in Madagascar; VITALIS, FEZU, and JGH University in Venezuela; and the minister of the environment in Yemen. These organizations and public representatives thus lent their stature in society to 350.org. By combining their events with others, as in Puerto Rico and Ann Arbor, Michigan, local organizers were able to link 350 with other causes. Though this had the potential to weaken the 350 message, it may also have provided a larger participant and audience base than that drawn by 350.org alone.

It should be noted that not all the events made use of local resources with respect to transmitting a message to policy makers. Some local organizers instead chose more intimate events for self-refection or to simply bring awareness to their own participants or a more local audience, as in the Vienna garden party and in Salzberg, Austria; in Ranobe, Madagascar; and in Denver and Puerto Rico. These events, with a focus on awareness or other environmental messages, alluded to a lack of unification on the 350 message and even regarding the actions required from the average citizen. For instance, messages tended to focus on stopping deforestation, buying organic or locally grown foods—rather than on participating in policy formation for climate change.

Resources chosen to amplify the 350 message tended to reflect local organizers' attempts to showcase community dynamics and interests in climate change. Though local announcements were publicized online and were thus open to anyone, either the type of event attracted a specific kind of participant or certain participants were explicitly invited to the event. Of the latter, youth served as the main target group for the purposes of encouraging policy makers to think of future generations, as demonstrated by the event announcements for Antananarivo, Madagascar; Catia la Mar and Maracaibo, Venezuela; and Yemen. Other events focused on general community participation, though events such as Denver, Colorado, United States, and Ranobe, Madagascar, may have been more exclusive owing either to the uniqueness of the event strategy (Denver's prayer flags) or to the specialized site (Ranobe's Spiny Forest). Like participants themselves, local organizers targeted audiences made up of their fellow citizens or their national policy makers, demonstrating a variety of agendas for such events and only cursory knowledge by some of how the public can gain attention in a political debate.

Local organizers' use of scientific knowledge as a resource could have perhaps been the strongest affirmation of public authority; they could have thus acknowledged their grasp of science and its place within the political debate. However, of the 17 events examined, only five made specific reference to such knowledge, and an additional event (Yemen) used imagery to infer a linkage between climate science and rising sea levels (assuming proper interpretation of image). Of these five events, all mentioned the significance of 350 as a goal for atmospheric CO_2 concentrations, though no elaborations on this number were made, at least in the announcements.

Unlike organizers' less obvious incorporation of rules into their rally announcements, their presentation and use of resources was more obvious and straightforward. This again speaks to an understanding on the part of local organizers about how to attempt to influence government officials without fully grasping the inner workings of these institutions.

THE POLITICAL SIGNIFICANCE OF 350.ORG EVENTS

350.org organizers labored under the premise that "if an international grassroots movement holds our leaders accountable to the latest climate science, we can start the global transformation we so desperately need" (350.org, 2011). In order to do this, they attempted to use public authority to influence the December 2009 UN climate negotiations, a strategy focused on the responsibility of leaders to their citizens by addressing their needs and demands. According to 350.org, this responsibility revolves around developing sound policies for reducing global CO_2 to levels safe for humanity. As with its Step It Up structure from the 2007 events (Endres, Sprain, & Peterson, 2009), 350.org continued the do-it-yourself event format whereby local organizers chose how to present their event to a greater public while incorporating the 350 message. This format attempts to teach event organizers how to insert the public into national deliberations on climate change and at the same time to give event organizers and participants freedom of expression in delivering the message. Like Step It Up, this led to a variety of interpretations on how to construct an event, including determining who would participate, how they would participate, who organizers considered their target audience, and how the event participants would communicate the 350 message.

Although local organizers demonstrated inspiration from 350.org by gathering on the preordained "day of climate action," the goals and focuses of the 17 local events analyzed did not always specifically match those of 350.org. Many announcements mentioned 350.org or the goal of 350 parts per million, but the direct goal and message of reducing worldwide atmospheric CO_2 was sometimes lost in local initiatives only indirectly related to controlling greenhouse gas emissions and climate change. Furthermore, the direction of the message was not always clear from the event announcements, or it diverged significantly from the 350.org rubric. Certainly, 350.org organizers wanted the message projected outward and focused on national leaders who could forge climate policy on international scales. Some events seemed intent on this; however, others focused internally on local communities or national issues.

This discrepancy between national organizers' and local event organizers' visions for the 350 rallies highlights an insufficient understanding on the part of event organizers regarding how to use societal structure to encourage public and policy-maker engagement with climate change. As exhibited in the Step It Up rallies, the do-it-yourself, loosely organized event format exhibited limitations, such as failing to make the movement visible to outside audiences and politicians, creating tensions and disconnects between participants and organizers, and leading to a lack of control by national organizers over local actions (Endres, Sprain, & Peterson, 2009)—all of which remained issues with the global

350.org rallies. More recent literature, therefore, has put forth a different approach to encourage action, a bridge between top-down and bottom-up approaches (Ockwell, Whitmarsh, & O'Neill 2009). This suggestion falls in line with Giddens's structuration theory in that it works within the confines of current institutions and takes advantage of shifting power in the form of public authority. It would also connect efforts on the part of climate policy experts *with* grassroots organizers as a means for unified action toward the creation of climate policies.

Even with the occasional divergences from and loose linkages to the 350.org message and purpose, local organizers demonstrated their ability to play with or subvert social rules and mobilize social resources in their events, thus enhancing public authority on anthropogenic climate change. This authority is dependent upon the ability of individual actors and collectives to affect the structure of society and its large-scale institutions (i.e., national governments and the UN) through a demonstration of global unification for change.

Conclusion

350.org sought to encourage national leaders to take responsibility for climate change and help set global policies for the reduction of greenhouse gas emissions. Through use of the Internet, 350.org was able to assist in the orchestration of 5,245 events in 181 countries, a spectacular feat of effort and organization, and one that fostered a strong sense of global unification that may be especially important given the relative failure of any officially sanctioned forward movement at the Copenhagen summit. These events generally took on identities of their own, serving local and national agendas with either obvious or somewhat loose connections to the 350 message. Overall, however, they were able to amplify climate change politics through mere affiliation with 350.org, allowing the NGO to use these numbers to demand attention from policy makers. In our examination of these events, we found Giddens's structuration theory a useful framework for understanding how the promotion of public authority may enable multiple possibilities for local residents to claim a voice in climate change politics. Using a structuration approach to gauge the value of 350.org encouraged us to recognize that all events may claim some level of success even as they retain their own identities and specific agendas for positive action on climate change and related social issues, including violence, environmental justice, healthy eating, job creation, and so on. Our analysis showed local event organizers' creativity in using and adapting rules and resources, as well as a lack of or inconsistent development of public authority across local events. By explicitly accounting for how communicative action and institutional structure interacted with both individual agents and social systems, a structuration approach to campaigns such as 350.org should help build an understanding of how new media and existing social structures can create spaces in which local residents can insert themselves into climate politics.

Part III

Public Engagement and Agonistic Politics

CHAPTER 8

ART AND POLITICAL CONTESTATION IN CLIMATE ISSUES

WHO OWNS THE AIR?[1]

Andrea Polli

> Economic superpowers have been as successful today in their disproportionate occupation of the atmosphere with carbon emissions as they were in their military occupation of the terrestrial world in colonial times.
> Andrew Simms, the New Economics Foundation (Simms, 2006)

INTRODUCTION

The Kyoto Protocol required 37 industrialized countries and the European Union (EU) to reduce greenhouse gas emissions by an average of 5% below 1990 levels by 2012. Although, or perhaps because, it is the world's biggest emitter of greenhouse gases, the United States is not a part of the protocol. In the Kyoto Protocol, the main "basket" of greenhouse gases to be reduced are carbon dioxide, methane, nitrous oxide, hydrofluorocarbons, perfluorocarbons, and sulfur hexafluoride (United Nations Framework Convention on Climate Change, 2007).

The Kyoto Protocol includes a global greenhouse gas emissions–trading system that has been in place in Europe since 2005. This emissions-trading system could also be called a cap-and-trade system. In the first year, credits are generous and the total amount of emissions for each company is determined. Each company has to achieve no more than that amount. Then, each subsequent year, the amount of credits allotted to each company is reduced, pressing companies to slowly lower emissions.

This cap-and-trade system, in essence, allows polluters to "own" parts of the atmosphere. Although this ownership might be optimistically considered a way for polluters to take responsibility, in the political arena, it amounts to a license to pollute without limit. This chapter examines models for owning ephemeral materials, how the concept of owning the air human beings breathe has become culturally acceptable, and the roles of art in that process. The chapter also explores how contemporary artists are attempting to challenge this acceptability by responding to the idea of air as a commodity.

In this chapter I examine how the concept of the carbon-trading market, or the buying and selling of air, has gained cultural acceptance, how works of art may have indirectly contributed to the idea's public acceptance, and how contemporary works of art are now challenging the ethics of such markets. First I outline several models for the carbon-trading market, including not only other emissions-trading systems but also the market for broadcasting rights. Next, I discuss cultural examples of buying and selling air, including those in the contemporary art movement referred to by Lucy Lippard (1973) as "the dematerialization of the art object." Finally, I examine several artists' works that challenge the market notion and promote other solutions to the anthropogenic climate change problem.

MODELS OF COMMONS TRADING SYSTEMS

In 1990 the United States launched a cap-and-trade program in sulfur dioxide as an amendment to the Clean Air Act. An initial objective of the program was to reduce sulfur dioxide emissions from utilities by 8.5 million tons below 1980 levels. To accomplish this, electric utility plants above a certain size were given initial emissions allowances based on historical patterns. The act included stiff penalties for excess emissions, at a value more than 10 times that of reduction costs. The program achieved a very high degree of compliance, and Congress considers the program a success (United States Congress, 1997). The current European Trading Scheme on greenhouse gases is modeled after the sulfur dioxide (SO_2) system, but it is not certain that this can work with carbon dioxide (CO_2).

Another model for carbon trading is the 1989 Montreal Protocol on Substances that Deplete the Ozone Layer, designed to protect the ozone layer by phasing out the production of a number of substances believed to be responsible for ozone depletion, in particular chlorofluorocarbons (CFCs) and halons. The Montreal Protocol created a system for the international trading of allowances. In the protocol, trading is combined with a tax to offset any large profits from allowances that might discourage the reduction of CFCs. Since the Montreal Protocol came into effect, the atmospheric concentrations of the most important CFCs and related chlorinated hydrocarbons have either leveled off or decreased. Kofi Annan, former secretary-general of the United Nations (UN), has maintained that the protocol is perhaps the single most successful international agreement to date ("Stratospheric Ozone Depletion," 1997).

However, by 1997 smuggling of CFCs from developing nations, where they were still permitted, into the United States and other developed nations had become big business. In Miami in 1997 CFC smuggling was believed to be second only to cocaine. Although the protocol's goal was to completely phase out these gases by 2000, as recently as 2005, several companies in eastern China were found to be involved in the illegal international trading of CFCs ("Corruption Stalls," 2005). A market solution can be very susceptible to this kind of exploitation.

The difference between a potential CO_2 market and both the CFC and SO_2 markets is that technology exists to clean up CFC and SO_2 emissions, and the customers of these two markets are only a relatively small number of companies with outdated technology. In contrast, the CO_2 system applies to thousands of companies, and despite optimistic talk of "clean coal" and carbon capture and storage, there is presently no technology to make coal and other fossil fuels burn significantly more cleanly or to sequester carbon emissions safely on a large scale.

Another model for distributing air rights that can be examined is the right of the airwaves, or broadcasting rights. Although broadcasting does not involve placing dangerous pollutants in the air, it does involve the corporate ownership and control of an ephemeral substance, a specific frequency that is transmitted through the air. The idea of broadcasting rights therefore may have played a role in public acceptance of corporate air ownership. The U.S. Federal Communications Commission (FCC), which was created by the Communications Act of 1934, regulates all nongovernment wire and wireless communications. The act specifies only that broadcasting be in the hands of American citizens and leaves it up to the FCC to decide how to license broadcast rights. In the past, applicants were required to describe plans for programming to be judged on general usefulness to the public and on its practicality. In practice, this resulted in a combination of private and governmental control.

However, since 1994 the FCC has conducted auctions of licenses for the electromagnetic spectrum, open to any eligible company or individual that submits an application and an upfront payment. According to the FCC, the auctions more effectively assign licenses than previously used hearings or lotteries did (FCC, 2007). This system created a broadcasting sphere widely dominated by the large commercial interests with the resources required to win this bidding war for the airwaves.

In 1996 the FCC relaxed the rules that restricted broadcasters from owning several radio or television stations in one market, allowing broadcasters, for example, to own an unlimited number of radio stations. This move has created an increase in advertising prices and caused an unexpected outcry from communities that have found that the consolidation of the media in the hands of a few large corporations has resulted in a loss of quality local programming; corporations instead "pollute" the airwaves with homogenized, computer-controlled broadcasts. The lesson to be learned here that might also apply to a market in greenhouse gases is that the deregulation of the market, allowing the unrestricted purchasing of air rights—whether to pollute or to broadcast—could well result in a disastrous reduction of the quality of air and the airwaves.

A fundamental aspect of the greenhouse gas emissions–trading system is the granting of property rights to the air. The idea of ecological economics, as illustrated in Garrett Hardin's (1968) essay *The Tragedy of the Commons*, came from the understanding that environmental resources are finite, and because these resources can be destroyed, incentives should exist for protecting them. The question of how these incentives should be created, however, whether through a trading system or through regulations or some other means, is a topic of great controversy. What follows are some of these conflicting viewpoints.

AIR AND PROPERTY RIGHTS

Ecological economics provides both a mechanism for the valuation of environmental resources and an incentive for keeping within an established environmental budget. In 1997 the U.S. Congress described it in this way:

> From an economic perspective, pollution problems are caused by a lack of clearly defined and enforced property rights. Smokestack emissions, for example, are deposited into the air because the air is often treated as a common good, available for all to use as they please, even as a disposal site. Not surprisingly, this apparently free good is overused. A primary and appropriate role for government in supporting the market economy is the definition and enforcement of property rights. Defining rights for use of the atmosphere, lakes, and rivers is critical to prevent their overuse. Once legal entitlement has been established, markets can be employed to exchange these rights as a means of improving economic efficiency.
>
> (U.S. Congress, 1997, p. 3)

The biggest beneficiaries of a greenhouse gas emissions–trading scheme can be found in the banking industry (in the United States, this industry has been lobbying heavily for the implementation of a carbon cap-and-trade system since Kyoto) and the nuclear power industry, which stands to gain from a potential loosening of restrictions on the production of new power plants (Gelinas, 2007).

However, despite the support of banks, not all economic experts favor the use of such a system. Many would rather see a carbon tax in place. In practice, a carbon tax functions very much like a trading system because polluting companies either pay a tax or pay for carbon credits for their emissions. Both systems also raise consumer prices on fossil fuels. However, critics of the tax system say that it does not provide the incentive or "race for the pot of gold" that the carbon-trading system instills by financially rewarding companies that can substantially cut emissions (Solomon, 2007).

There are critics of the trading system on the political right and left alike. Arguments for and against this system range from concern about flaws and possible abuses of the system to criticism of its fundamental assumptions about ownership. From the right, Nicole Gelinas, a journalist of *The Wall Street Journal*, has argued against a trading system in the United States, stating that the global competition created by the system will hurt U.S.-based corporations. If U.S. energy companies cannot limit emissions, the international cap-and-trade system would require them to buy emissions credits from other countries. Interested in protecting the financial interests of these U.S. corporations, she called the trading system a direct subsidy to developing nations that would essentially pay for their power-plant upgrades (Gelinas, 2007). She argued that a federal carbon tax, an alternative to the international carbon cap-and-trade system, would provide revenue to the U.S. government that could then be used to subsidize U.S. power-plant upgrades, whereas an international cap-and-trade system could put this revenue in the hands of other countries, particularly those that have already reduced emissions (Gelinas, 2007). Gelinas's views seem to align with those of the U.S. government. In fact, one of the stated reasons the United States has opposed the Kyoto Protocol is that it provides exceptions for developing countries, the countries that stand to benefit most from the protocol and lose the most from the consequences of global warming.

In *The Weather Makers*, Tim Flannery debated this position, stating that although developing nations were not bound by the Montreal Protocol, it was very successful. Flannery had mixed feelings about the subject of carbon trading. On the one hand, he observed that emissions trading is cost effective, and that as a tool to reduce pollution it has been successful in the past (for example, in the case of SO_2 trading). On the other hand, Flannery wrote about the case of Eastern European countries whose economies have suffered ruin since the 1990s and therefore are producing approximately 25% less CO_2 than they were in 1990. Because the Kyoto trading system requires emissions to fall by only 8% of 1990 levels, without doing anything to actively lower emissions, these countries end up with a surplus of carbon credits, known as "hot air" to critics of the protocol. Flannery posited that creating a new global currency is too risky, for the foundation of any currency is based on trust—in this case, trust that the seller will lower emissions—and so far, he has not seen any guarantee that this will happen (Flannery, 2005, p. 225).

Arguing from the political left in *Carbon Trading*, Larry Lohman (2006) quoted Flannery's work when outlining the potential effects of climate change, but he was much more pessimistic than Flannery, particularly in regard to the commercial markets. He focused on the seriousness of the problem and predicted that many markets will eventually collapse. He used the example of the insurance business, providing quotes from insurance specialists who estimate that chaotic climate change could increase insurance rates several times the rate of world economic growth, creating a situation where the world economy would not be able to sustain the losses and would collapse.

Lohman was critical of the hopeful attitude held by many industrialized countries that foresees technological developments—such as carbon sequestration—that would allow continued use of fossil fuels as the solution. He devoted a chapter to analyzing these technological solutions and concluded that they are nothing but a smokescreen used to distract public attention from the government's lack of necessary action. He called this distraction the second strategy; what he called the first strategy is the public denial of the existence of anthropogenic climate change.

> The *first strategy* works to reshape or suppress understanding of the climate problem so that public reaction to it will present less of a political threat to corporations. The *second strategy* appeals to technological fixes as a way of bypassing the debate over fossil fuels while helping spur innovations that can serve as new sources of profit.
> (Lohman, 2006, p. 34)

In *Earth in the Balance*, Al Gore embraced in part what Lohman critically called the second strategy, proposing a strategic environment initiative (SEI) that like the existing U.S. strategic defense initiative (SDI), would require a major national effort, but with a focus on the environment rather than the military. Although he did emphasize that any new technologies should be evaluated, his focus was primarily on developing new technologies to combat climate change (Gore, 2006).

Finally, Lohman was very critical of what he called the third strategy, primarily because of the property-rights issue implicated by privatization of the air: "The *third strategy* appeals to a 'market fix' that secures the property rights of heavy Northern fossil fuel users over the world's carbon-absorbing capacity while creating new opportunities for corporate profit through trade" (Lohman, 2006, p. 34). Like Flannery, he was concerned that if the top-down creation of a greenhouse gas–emissions market happens without public debate, it will result in a market filled with distrust, including a lack of faith in both the structure and the implementation of the market. The bottom line, according to Lohman, is that the current emissions-trading structure allocates the most allowances to the biggest emitters, effectively giving a handout of billions of dollars to the most egregious polluters and providing them with incentives to continue polluting.

In *The Great Emissions Rights Giveaway* (2006), Andrew Simms, policy director of the New Economics Foundation (NEF) and the Foundation for the Economics of Sustainability (Feasta), proposed an alternative structure for EU carbon trading. In Simms's proposal, the EU's emissions allowances would be divided on an equal per capita basis and distributed to every EU resident. Residents would then be able to sell these allowances to companies or keep them off the market, thus promoting cleaner air. Simms compared this approach to other alternative approaches—for example, David Fleming's proposed tradable energy quotas (Simms, 2006), which allow governments to provide each citizen with a portion of carbon units equal to the amount of the general public's fossil fuel use. Each citizen can use these units to purchase energy and fuel or sell unused units on an open market. Carbon units used by industry are sold at an auction similar to the FCC broadcast-rights auction. In another example, the Sky Trust in the United States proposed a system in which money made from the sale of emissions credits is kept in a trust in order to ensure that any financial gains are used to benefit the public. The proposals of both the Sky Trust and NEF/Feasta are based on the fundamental principle that the atmosphere belongs to all people equally and not to any governments or corporations (Simms, 2006). Although there are strong voices that argue against the carbon-trading system, outrage about the idea of selling air has not been prominent in the public discourse. Dominated by Gore's market-fix idea, discussions have focused on the effectiveness of strategies rather than their underlying ethical implications.

Air for sale in contemporary art

Drawing on Noel Castree's (2003) critique of the commodification of nature and Thornes and Randalls's (2007) discussion of how this process applies to the global atmosphere, I refer to six examples of the ways in which the atmosphere has been commodified: as fuel for the combustion process; as a packaged industrial gas product, such as nitrogen; as value added to property owing to air quality or good climate; as a site for geoengineering services, such as cloud seeding for weather modification; as the cause of problems that require health services, such as lung transplants; and as weather and climate data and information services.

One might think that the idea of air for sale is absurd. After all, people would not *actually* pay for the air they breathe. Culturally, however, this may no longer be the case, as evidenced by the rising popularity of the oxygen bar and portable canned air. In both cases, oxygen is touted as a cleansing medical therapy. Advertisements focus on the healing power of air and of oxygen therapy. Advertisers promote air as energizing for exercise, effective in combating cigarette smoke and curing a hangover. They sell air based on the idea of its being pure, fresh, and clean, and many promote it as an escape from the smog of city life. In the case of the oxygen bar, customers pay for a session of five minutes or so,

Art and Political Contestation in Climate Issues 231

during which they are able to relax and breathe clean, sometimes scented air. The oxygen bar started as a trend in the 1990s in Japan, Mexico, and South America and quickly spread to nightclubs, spas, casinos, and malls in Europe and the United States. In 2003 the oxygen bar at Olio!, a restaurant at the MGM Grand Hotel in Las Vegas, boasted 200 to 400 customers per day (Rimbach, 2003). Portable canned air is becoming just as popular and widespread. In Japan a recent large-scale commercial venture is O2supli, a portable can of oxygen.

> The idea behind the product is to allow buyers to replenish their oxygen levels anytime they feel a lack of it due to stress, fatigue, or other factors ... Each can contains enough oxygen for 35 two-second inhalations, meaning each can lasts for roughly a week if it is used five or six times a day. At first the canned oxygen will be sold in Tokyo ... then at all 11,000 of Seven-Eleven Japan's nationwide stores.
> ("Sales of Canned Oxygen," 2006)

The popularity of the oxygen bar and canned oxygen prompts a question regarding how global culture has reached a point in which the absurd notion of buying and selling the air is acceptable at any level, corporate or individual. More specifically, this chapter discusses the roles the arts have played in making the commercialization of air a culturally acceptable activity and speculates on how those roles may be altered.

Artists in the conceptual art movement in the 1950s and 1960s made the radical proposal that air, wind, temperature, and weather could actually be art materials in themselves. Yves Klein's concept of *air architecture* challenged the definitions of art and architecture by building with ephemeral materials, but it may have inadvertently contributed to the commodification of air. Klein was interested in the ways that human beings can use science and technology to conquer the ephemeral, to the point of turning even air and fire into building materials. Klein saw science and technology as the saviors of architecture because they promoted new forms and structures made from sculpting the air and other "immaterial materials." He believed that air architecture would actually improve the environment, positing that air architecture must be adapted to the natural conditions and situations, to the mountains, valleys, monsoons, and so on, if possible without requiring the use of great artificial modifications (Noever & Perrin, 2004). By creating air architecture, Klein radically redefined the limits of architecture, asking why solid materials like stone, wood, and brick should be the only materials in an architect's palette. By developing technologies for using (or some may say, misusing) ephemeral materials like air, water, and fire for architecture, Klein transformed the architectural imagination.

In the late 1960s, a group of artists, including Robert Barry, became associated with art dealer Seth Siegelaub and started producing work that questioned the limits of art. Barry's series of works known as "invisible" art included *Inert Gas Series* (1969), in which a specific amount of the gases neon, xenon, and helium was released from measured volume to indefinite expansion in the Mojave Desert (Wood, 2002, pp. 35–36).

Lucy Lippard observed in *Six Years: The Dematerialization of the Art Object* (1973) that novelty is the fuel of the art market, and at the time of Robert Barry's *Inert Gas Series*, this fuel was being burned at a rapid pace, constantly stretching the boundaries of the definition of art. By using the term *dematerialization*, Lippard attempted to remove art from its status as commodity by creating such ephemeral objects of art as those including the natural expansion of gas, a substance it would be absurd to commodify. Artists like Barry were reacting against and attempting to create an alternative to the art market. However, over time, this work has not been removed from that market. The ephemeral art movement was actually embraced by commercial galleries and dealers. As Lippard and Chandler (1968, p. 31) suggested, this market needed the fuel of these increasingly novel ideas.

A connection to the market may have been necessary for the creation of the art and the survival of the artist; nonetheless, it created a paradoxical situation in which the immaterial was moved into the material realm.

Art and Political Contestation in Climate Issues 233

The critical stance of the artist on the art market was compromised through the positioning of the work within the art market. Although these immaterial works could be considered *critical* art as defined by political theorist Chantal Mouffe (2006) because they contribute to questioning the dominant hegemony of the art market, over time this function has been diminished because many ephemeral works have been commodified. Contemporary artists have experimented with ways to create critical art that addresses market dominance more directly and that may avoid this paradox.

In 2005 Tue Greenfort made *Bonaqua Condensation Cube* as a homage to Hans Haacke's 1963 *Condensation Cube*. Using Bonaqua, a popular brand of bottled water, as the water of condensation was meant to directly address the issue of ownership—both of the natural resource water and of an ephemeral artwork. Water was considered a public resource in 1963; however, by the early 21st century, it had become a commercial product. Like the earlier work, the piece was positioned in a gallery as an artwork to which a monetary value could be attributed. Like the earlier work from Haacke, *Bonaqua Condensation Cube* satirized the absurdity of the art market, but unfortunately their exhibition in a gallery has positioned both works as problematic parts of that commercial system.

A recent public work related to air was Laurie Palmer's *Hays Woods/ Oxygen Bar* (2005). This project highlighted the natural processes that create air and the idea of air as a public resource. Palmer used the format of the oxygen bar, but she subverted the commercial concept of the bar by providing the air for free. Customers at the *Hays Woods/Oxygen Bar* learned that the oxygen was actually coming from the same plants that surrounded them in the woods and realized the value of that green space. In her statement for the work, Palmer used the language of advertising:

> The oxygen bar is a mobile breathing machine, offering free hits of "natural" oxygen on a first-come, first-serve basis. This oxygen is produced by the photosynthetic work of green plants (from

Hays Woods) and is offered as a public service. It reproduces in miniature the beneficial cleansing and refreshing effects of city green spaces on the air we breathe. The oxygen bar anticipates the imminent loss of public resources that filter Pittsburgh's dirty air and replenish it with oxygen, in particular, Hays Woods. At the same time, the oxygen bar anticipates the active participation of citizens of Allegheny County in land use decisions affecting our public health.

(Palmer, 2005, n.p.)

As evidenced in this quote, Palmer had a specific political goal in presenting this work. She wanted to encourage active participation of citizens in land-use decisions. Palmer's work used the metaphor of an established commercial product (oxygen) and the language of advertising related to health and well-being in an attempt to convince her audience to appreciate the value of the natural setting of the woods and promote preservation. This work creates an alternative pathway in advocating for public action. It is a creative and critical intervention that reaches new audiences with an alternative viewpoint. As scholar Malcolm Miles (2010) has explored, intervention as art in the realm of climate change may have a capacity to contribute to a shift in public consciousness.

THE *AER* EXHIBITION AS POLITICAL INTERVENTION

In the 21st century a growing number of contemporary art and social media projects from a multidisciplinary perspective represent the beginning of an effort to give a voice to each person affected by poor air quality, to highlight the flaws in the current system, and to empower individuals to preserve and protect the earth's fragile atmosphere.

I curated the *Aer* exhibition as part of my research; the exhibit featured artists whose works are far from neutral about the issue of air and air quality (Polli, 2008). The artists in *Aer* all look critically at air issues and use various methods to raise awareness among the public. Some also take an active role by directly affecting the air quality and therefore human lives. Because air is invisible, artists are faced with the challenge of making the intangible tangible. Air pollution is a silent killer, and it is challenging to give a voice to the body's dependence on clean air. Most of the featured projects blur the line between art and activism, and all the projects aim to change the public understanding of air as well as to question accepted norms of ownership and responsibility.

Climate change is a global problem that requires international cooperation. The projects in the *Aer* exhibition navigate the economic and personal politics of air and air quality. Because these politics can be complex and controversial, Platform London, a group whose work crosses disciplinary lines to achieve social and ecological justice, attempts to blur the boundaries between art and activism. The group's projects involve rigorous research, advocacy, public art, and education, or various combinations of these. The global price of oil is set at London's International Petroleum Exchange, and the artwork titled *Unravelling the Carbon Web* looks closely at two major players in the oil industry that have headquarters in London, BP and Shell. The project focuses on their activities in Iraq, the former Soviet Union, and Nigeria—including the Nigerian oil industry and Shell's policy there. A component of the project by Platform London, the *Remember Ken Saro-Wiwa* project, was a tribute to the Nigerian writer and activist who led a nonviolent campaign against the environmental damage associated with the operations of Shell and other multinational oil companies in the Niger Delta. His execution by the Nigerian military in 1995 provoked international outrage. Platform London then coordinated a coalition of organizations and individuals to contribute a series of living memorials to Saro-Wiwa, which included a book of poetry and a stainless steel bus made by the Nigerian-born artist Sokari Douglas Camp. This growing project, aptly named *Unravelling the Carbon Web*, contains an archive of related news, historical documents, and analysis, including fables that try to reach the conscience of the global oil industry.

Other works featured in the *Aer* exhibition are Los Angeles–based artist Kim Abeles's *Smog Collectors* series, which makes the invisible visible by using the air as an almost photographic medium, placing material surfaces on her rooftop and allowing particulate matter to collect on these surfaces over time. The resulting images look like beautiful photograms, but they present the shocking effects of particulate pollution. The *Presidential Commemorative Smog Plates* were also made as part of this series. Abeles exposed each plate to smog for amounts of time that corresponded to the environmental record of each U.S. president. This process caused the resulting images on the plates to darken with each successive president, strongly illustrating the tragic decline in U.S. air quality with each new presidential administration. Abeles used a playful method of illustration as an alternative pathway for informing the public about the relationship of governmental policies to air quality.

Another project that uses the visualization of the air to make a powerful statement is *Pollstream*, a series by Hehe (Helen Evans and Heiko Hansen). Using interactive media and sophisticated visualization of the composition of air and smoke, these works inform and alert the public to real-time local air quality. Monet found the skies of European cities awash with unusual colors caused by human-made pollution. In a similar way, Hehe's *Champs D'Ozone* overlays a real-time image of Paris skies with colors representing the unseen pollutants contained within. On one level, Hehe's *Pollstream* projects pay homage to Monet's 19th-century works by aestheticizing the air and smoke, but on another level, unlike viewers of Monet's paintings, *Champs D'Ozone* viewers can directly link the colors to actual levels of pollutants. Hehe's work forces viewers to pay critical attention to what the colors represent, a deteriorating quality of air.

Another elegant example of local air visualization is Sabrina Raaf's *Translator II: Grower*. In this project, a tiny robotic rover drew a simple green line at the bottom of a white wall, beginning parallel to the floor. This line indicated the level of CO_2 in the room: the taller the line, the more CO_2. The rover moved around the room creating a horizon of tall and short grasses, a history of the changing airscape. This interactive work responds directly to the number of people in the room, for all human beings exhale CO_2.

Australian performance artist Sarah Jane Pell has also highlighted the body's transfer of air and people's dependence on air as living, breathing beings. Her works explore the physical and emotional limits of the body. For example, *Interdepend* creates a closed-circuit life-support system between Pell and artist Martyn Coutis, and *Undercurrent* presents a single performer contained within a sealed transparent dome with a finite amount of breathable air. These works are physically demanding for the performers and create an overwhelming emotional intensity for the public. In an important historical political document outlining the effects of industry on air quality and public health, *Fumifugium* (1661), John Evelyn referred to the air as the soul or spirit of humankind, and Pell's works seem to give that soul or spirit a physical manifestation. This can be witnessed either through human interdependence or through a single womb-like containment that without breathable air would become a tomb. Like Raaf's work, Pell's works seem to hold a vision of the future. For example, in *Translator II: Grower*, Raaf presents a robot that methodically records the human imprint on air. This rover, which can continue to perform its duties even when the air becomes toxic to humans, questions humanity's future on the planet if people continue to poison the air. In Pell's case, this vision is apocalyptic: one in which the very air human beings breathe is a limited commodity. In the future, will the earth's fragile atmosphere continue to sustain humankind, or will people be forced to remain contained in controlled environments while machines roam freely, reporting on the world outside? By suggesting an alternative future of limited air, these works by Raaf and Pell create a pathway that forces audiences to consider the important reciprocal relationship between humans and the earth's atmosphere.

Evelyn (1661) called air the "vehicle of the soul," and the term *vehicle* became a metaphor for defining air in the context of the *Aer* exhibition. Many of the various definitions of *vehicle* fit not only the specific artworks in *Aer* but also many contemporary artworks. For example, a vehicle can be a means of transmission and a medium of communication, a means of accomplishing a purpose, and also an idea to which the subject of a metaphor is compared. A vehicle can be a means by which an idea is transmitted, through some kind of tangible art medium like a painting or sculpture, yet in the current artistic climate (the age of the dematerialism of the art object), air has become a viable art medium. Although air is as fleeting as an idea, and perhaps even for this reason, it can also be the vehicle through which an idea is expressed, as shown through the work of Yves Klein and Robert Barry.

The artistic team called Superflex has created a project that contained gas for the purpose of influencing social change. Superflex worked in collaboration with Danish and African engineers on *Supergas*, which aims to provide a modest and efficient portable biogas system for families in Africa. Superflex has identified its artworks as tools shaped by a social and economic commitment. The tool *Supergas* has helped individual families become independent producers of energy with minimal time investment and without making dramatic changes to their culture. This transdisciplinary process of development involved the community on a deep level. Moreover, the design process did not end with implementation; the tool itself was originally designed for flexible use, a kind of open-ended platform. *Supergas* captures organic materials that are an ordinary artifact of farming in order to promote individual energy empowerment and to protect the air from more polluting forms of energy production. The thought process that with minimal costs turns a polluting compound into an energy resource can be a contagious activity, in turn promoting similar independent and innovative cross-disciplinary and cross-cultural cooperation in the future.

Art and Political Contestation in Climate Issues 239

In the 1960s the United States first implemented air-quality monitoring and advisory systems to alert the public about dangerous levels of pollution. During days of poor air quality, public notification came bundled with the weather report, and people were told to avoid going outside. Although this one-to-many information exchange may work effectively for this kind of warnings, other models in a networked environment could be exploited with respect to air quality. In an open platform, *Area's Immediate Reading* (AIR) by Preemptive Media, artists Beatriz da Costa, Jamie Schulte, and Brooke Singer turned individual citizens into volunteer smog detectives using a network of wireless pollution-monitoring devices. This open platform allowed real-time sharing of location and time-based information about pollution, health, and the environment. The media used by artists in this exhibition became a strategy for opening public dialogue. This wireless network created for the *AIR* project functioned metaphorically as another definition of a vehicle: a medium in which medicine is administered. In this metaphor, the illness is a complacent society, blind to the dangerous effects of air pollution, and the medicine is increased public participation in air-quality monitoring.

Amy Balkin's *Public Smog* (2007) project addressed the commodification of clean air from the perspective of the market. Balkin is an artist whose works create metaphoric shifts, question social and political assumptions, and take an activist stance. Her work often involves intensive legal, financial, and political research. *Public Smog* is a public park located in the atmosphere; it is of changing size and floats in an unfixed location. At first it seems like a fantasy. How can anything exist in such an ephemeral location? The premise of Amy Balkin's work is to engage the economic system of emissions trading. Through her project, the global public can purchase as many emissions allowances as possible on the emissions market. These carbon offsets are then retired—in other words, they are taken off the market—making them unavailable to polluting corporations (figure 6). By openly embracing the free market for the public good, Balkin has presented a sharp critique of the system that the project must operate within. Therefore she sees her public space as the emissions-trading market.

Balkin's work clearly questions the market, and the solution she proposes is meant to be absurd. However, in the context of contemporary culture, the solution seems like a viable one. In fact, it is very similar to the structure proposed by Feasta in which individuals buy and sell emissions credits on the market and a certain amount of emissions credits are distributed for free to citizens. A group called TheCompensators* proposed the exact same solution Balkin did, claiming to have retired EU emissions allowances equivalent to 100 tons of CO_2 as of November 2011 (TheCompensators*, 2011). These solutions create potential problems grounded in the very idea of the market. Healthy markets may grow, but if people decide to buy emissions credits or clean air, the market may issue more emissions credits to balance the market and meet the demand. This process would force the public to pay ever-higher prices for clean air.

The difference between Balkin's *Public Smog* and TheCompensators* project lies in the fact that one identifies itself as an art project and the other as an environmental project—but the underlying metaphors are similar. By using the metaphor of a public park, Balkin's work allows viewers and participants to look at the system of emissions trading through a familiar lens. Most viewers understand the difference between public and private property in the context of land use. The air as a public park is a metaphor Balkin used to influence greater understanding of the problems that may emerge from the privatization of the atmosphere.

The message of *Public Smog* becomes even clearer when seen in relation to one of Balkin's earlier works entitled: *This is the Public Domain.* In 2003 Balkin purchased a 2.5 acre parcel of land located in Tehachapi, California, intended as a permanent, international commons, free to everyone, held in perpetuity.

Figure 6. Public Smog Offsets Tomorrow Today (Douala, Cameroon, 2009).

Source. Photograph by Benoît Mangin. Reproduced with the photographer's permission.
Note. "A series of thirty billboards installed across Douala, Cameroon, prefigure the possible opening of a clean-air park over Africa. The billboards combined images of the local landscape with fifteen slogans in English and French, interweaving rhetoric, boosterism, greenwash, and agitation" (Balkin, 2012).

Because there was no precedent in the creation of such a commons in the current legal system, Balkin's project involved a complicated legal process that explored solutions in both real property law and copyright law. Through this process, Balkin questioned the foundations of the existing laws and highlighted the disappearance of public space. Balkin has identified the project as an art project, and this classification is essential to her legal approach, one that explores the land as a creative work of art. Through this project, she has joined a movement of people who move fluidly among the roles of artist, environmentalist, and activist.

On December 11, 1997, the Kyoto Protocol was adopted by the United States and 121 other nations, but it has never been ratified by the U.S. Congress. Influential leaders within the U.S. industrial sector have lobbied against ratification, predicting disaster if CO_2 reductions are enforced. In March 2001 the United States again pulled out of Kyoto, saying that complying would destroy the economy. Many U.S. citizens disagree with their government's actions and are concerned about rapid increases in global temperatures, melting ice, and rising sea levels—changes that the 2007 UN intergovernmental report on climate change states are unequivocal (Intergovernmental Panel on Climate Change, 2007a). In another example of contemporary artists' responding to the politics of air, Ben Engebreth provided individuals in various U.S. cities the chance to personally comply with the Kyoto Protocol. Engebreth has not identified himself purely as an artist, nor has he called *Personal Kyoto* an art project, and his projects often operate outside of the art market on the Internet. *Personal Kyoto*, also featured in the *Aer* exhibition, analyzes electric usage information and calculates an energy-reduction goal of what the Kyoto Protocol might require from an individual. *Personal Kyoto* allows individuals to monitor electric use on a daily basis with the goal of reducing their personal consumption of greenhouse gases. Like Balkin's *Public Smog*, *Personal Kyoto* works within an existing system to empower public action and benefit. Whereas *Public Smog* takes on the publicly traded carbon-offset system with collective action using a public space model, *Personal Kyoto* looks to individual responsibility and accountability as a means to encourage global change.

Projects that focus on personal responsibility embrace the philosophy of voluntary emissions reduction on an individual scale. Though these projects may help raise awareness of the problem of CO_2 emissions among the general public and thus may have an impact on total emissions, their focus on personal choice may detract from the urgent need to curb emissions now. Can the American individualist ideal (the ideal behind the problematic proposal made by the United States to the EU that advanced voluntary emissions reduction) promote the changes needed?

Voluntary emissions reductions may be a great idea on an individual level, but is this enough to mitigate the climate change catastrophe?

The paradoxical problems and metaphors raised by these artworks are part of the system in which the works exist. They are either in the art world, with its gallery economy based on the buying and selling of works, or in the public art world, in which works can be supported by government or private interests. Some works also operate in the semi-public forums of the emissions-trading market or the Internet. These works bring up larger questions about the potential of art in a time of global environmental crisis, and more specifically, about the potential of art and science collaboration in creating alternative pathways to understanding and responding to climate change.

Closing remarks

The work of the artists in my curated exhibition *Aer* and the related art and activist projects discussed in this chapter represent the beginnings of an effort to address the issues of air pollution and climate change from a multidisciplinary perspective. In the 1960s Yves Klein, Robert Barry, Hans Haacke, and others developed artworks from ephemeral materials. By making artworks from air and water, these artists pushed boundaries and expanded the definition of art. At the time that Klein, Haacke, and Barry were creating the ephemeral works discussed in this article, water and air were seen as public resources. However, by the early 1990s bottled water had become a popular item in stores, and air pollution–trading systems in SO_2 and CFCs had already been implemented. Because artworks generally operate in various economies, whether they are gallery works or works of public art, the position of ephemeral artworks as private property may have unintentionally aligned with the idea of selling bottled water and the idea of buying and selling air. Though the public audience for a high-end conceptual artwork may be very limited, artworks have often influenced advertisers looking for enigmatic images that promote style and sophistication, bringing these ideas to the mass public.

Recent contemporary artists using ephemeral materials—for example, Laurie Palmer (*Hays Woods/Oxygen Bar*) and Tue Greenfort (*Bonaqua Condensation Cube*)—have attempted to challenge the idea of selling air and water by offering an alternative reading to the mainstream commercial messages that encourage marketing these resources. By presenting the *Hays Woods/Oxygen Bar* as a public artwork in which clean air is distributed for free, Palmer developed an alternative mechanism to access new audiences for art and to encourage social activism in support of preserving the woods.

Contemporary artists presented in this article have used dramatic visuals in order to offer alternative viewpoints. Both Kim Abeles (*Smog Collectors*) and Hehe (*Pollstream* and *Champs D'Ozone*) developed innovative ways to visualize air quality. Their works make the invisible tangible to art audiences in dramatic ways. Other artists presented in this article have used technology as a means of creative and critical intervention. Sabrina Raaf (*Translator II: Grower*) and Sarah Jane Pell (*Interdepend* and *Undercurrent*) have designed works that highlight the human body's impact and dependence upon air. Their works suggest the implications of an alternative future in which air is a limited resource.

As demonstrated in the controversies surrounding the creation of global markets that trade in air and through artists' critical responses to this controversy, there can be new mechanisms for communicating problems related to the environment in general and climate change in particular. From a top-down perspective, carbon-trading systems may put the atmosphere under the control of a small number of large corporations rather than in the hands of the citizens who need to breathe air to survive. However, artists may be able to play a role in popularizing another kind of system, one that facilitates a grassroots exchange of ideas in opposition to corporate control.

Art and Political Contestation in Climate Issues 245

Toward that aim, *Supergas* operates not only outside the traditional art market but also outside the global energy market. *Supergas* functions as a catalyst for social change by turning consumers into producers. Engebreth (*Personal Kyoto*) drew attention to global efforts to reduce greenhouse gases and also redirected the power of that reduction to individual citizens. Preemptive Media (*AIR*) also placed tools in the hands of citizens by empowering the public to record and monitor local air quality, and Amy Balkin (*Public Smog*) critically questioned the carbon-trading system by suggesting a way for individuals to collectively subvert the system and return the air to the public domain.

Each human being is affected by poor air quality and climate change, and the response of governments to privatize the air may not solve the problem but may rather create more difficulties. Several artists discussed here have created alternative pathways, highlighting the flaws in the existing trading systems and empowering individuals to act.

In his analysis of "cultural climatology," John Thornes (2008) pointed to a long history of artistic responses to weather and climate science. Increasingly, contemporary artists are becoming involved in the *politics* of climate change by challenging hegemonic constructs, such as carbon-trading systems, and by troubling the value systems underpinning them. Processes of the reception and consumption of artistic projects on the commons and on climate change lie beyond the scope of this chapter but would be fertile terrain for future research. In any case, as this chapter has shown, "critical art" (Mouffe, 2007) can serve as an alternative medium for promoting activism, as well as new social interactions in the public space that have the potential to transform political engagement with climate change.

Endnotes

1. This chapter is adapted from a presentation given at the 11th annual convention of the Media Ecology Association, University of Maine, 2010. See *Proceedings of the Media Ecology Association*, Volume 11, 2010.

CHAPTER 9

VISIONS OF CLIMATE POLITICS IN ALTERNATIVE MEDIA

Shane Gunster[1]

CLIMATE POLITICS, POLITICAL AGENCY, AND NEWS MEDIA

The intersection of climate change communication with the *politics* of climate change, as distinguished from the representation and communication of climate science, has begun to attract much greater attention from critical scholars and activists. One of the foremost North American experts in climate change communication, Susanne Moser, has recently called upon climate change communicators to "go beyond merely conveying climate change knowledge and more effectively encourage and enable individuals to take part in the societal transformation necessary to address climate change successfully" (2009, p. 284). Taking aim at what he perceives as the top-down, professionalized, marketing approach favored by some advocacy groups and communication professionals, Robert Brulle argued for an activist reorientation of environmental communication away from identity campaigns in favor of a renewed focus on civic engagement through building up "alternative field frames" that create "an alternative map of the social world around which individuals can collectively mobilize" (2010, p. 85). David Ockwell, Lorraine Whitmarsh, and Saffron O'Neill likewise suggested that rather than promoting individual behavioral change, climate change communication should focus on stimulating social and political demand for climate regulation, motivating the public to exert direct pressure upon governments to enact more aggressive mitigation measures (2009, p. 320). As widespread criticism of the information-deficit approach to environmental communication suggests, simply providing more or better information about climate change (to counteract the effects of climate skeptics' promoting doubts about the level of scientific consensus, for example) will not be enough to motivate political mobilization. Instead, far more attention must be given to the specific ways in which different ways of communicating (or not communicating) about climate politics can either inspire or inhibit public engagement with climate change *as a political issue.*

As Anabela Carvalho persuasively argued, such engagement is invariably mediated by the representations of climate politics one finds in the news media. Media, she wrote,

> are the main arenas for citizens' understanding of political struggles in our times.... In a media-saturated environment, perceptions of the distribution of power, of the role of individuals in democracy and of the effectiveness of civic action are a function of multiple discursive representations.... [M]edia(ted) discourses also influence people's view of their own position in the chessboard of politics and are also constitutive of the political self, cultivating dispositions to action or inaction.
>
> (Carvalho, 2010, p. 174)

Visions of Climate Politics in Alternative Media 249

Portraying climate politics and policies as the exclusive domain of national and global elites, for example, leaves little space for individuals to engage as anything but powerless, cynical, and apathetic spectators. Conversely, descriptions that emphasize the potential for citizens to actively shape climate policies through participation in the political process and pressuring governments can foster much greater perceptions of political self-efficacy and, consequently, heightened levels of political engagement.

Beyond portraying the action (or inaction) of citizens in particular ways, the representation of government policies, programs, and regulations is equally important in framing the efficacy and desirability of political activism: Success stories about specific interventions that have spurred the development of renewable energy supplies or reduced greenhouse gas emissions are likely to strengthen a more general perception of *all* governments as, at minimum, hypothetically capable of effective action and thus worthwhile sites of political struggle and contestation. However, the absence of such accounts, coupled with an excessive emphasis upon the failures of government, can easily reinforce the opposite perspective—namely, the idea that governments are not only unwilling to take but also essentially incapable of taking meaningful action. Once government disappears as a site of (potential) political agency and individuals lose the belief that they (or those like them) could ever influence the conduct of political power, civic engagement—indeed, the very ideal and practice of democracy—can only ever appear as a pointless exercise in self-deception.

A recent survey of regional Canadian media coverage of Copenhagen (Gunster, 2011) identified fundamental differences between commercial/corporate media on the one hand and alternative/independent media on the other in terms of their representations of climate politics. In particular, the latter offered a much more diverse and optimistic vision of climate politics as a place in which broad civic engagement on climate change can challenge and overcome institutional inertia and can model democratic and participatory approaches to the development of climate policy. Despite the obvious affinities between calls for increased political engagement on climate change and alternative media that furnish ideal sites to nurture and promote such engagement, alternative media have received virtually no sustained attention from scholars of climate change communication.[2] Instead, the priority for most researchers has been documenting the deficiencies and limitations of large, national mainstream media, along with developing recommendations for improving environmental communication.

Though such work has been extremely useful and influential, it is equally important to explore existing media institutions and practices that *are* communicating about climate change in an effective and engaged manner. Just as success stories about (some) governments getting climate policy right can invigorate one's sense of political efficacy, success stories about (some) media getting climate politics right can affirm one's sense of how effective news media could be in motivating broader civic engagement with climate change. Identifying best media practices can also sharpen the critique of mainstream media insofar as it provides concrete evidence that a more radical approach to environmental journalism is not simply idealistic speculation but is, rather, already being vigorously practiced. Finally, a better appreciation of the virtues of alternative media by climate change communicators and scholars can help avoid the "reinventing the wheel" that often accompanies criticism of commercial media. Instead of pouring all the conceptual and logistical resources into developing new institutions, venues, and practices of climate change communication, it may be more effective to plug into existing media structures that resonate with many of the principles suggested by critical scholars and communication practitioners.

"Future research," wrote Carvalho, "should focus on the role of new and alternative media ... in the empowerment and performance of resistance in climate change-related movements. It should analyze the communicative spaces where new forms of political subjectivity may be developing. It should probe into new varieties of 'collective political subjectivity' which are likely to be multi-scalar and flexible" (2010, p. 176). Mindful of this need to broaden the approach to studying media and climate change, I undertook a systematic review of the ways independent, Vancouver-based Canadian media organizations—*The Georgia Straight* and *The Tyee*—covered the politics of climate change during 2009. As Canada's largest free alternative urban newspaper with both a weekly print edition (weekly circulation, 120,000; weekly readership, 804,000; "The Georgia Straight," 2010) and online content, *The Georgia Straight* offers readers a combination of independent news, opinion, lifestyle, and entertainment features. *The Tyee* describes itself as "Canada's leading independent online news magazine" and publishes news and opinion features, as well as an associated blog, *The Hook*, that are accessed by more than 150,000 readers per month ("The Tyee," 2010). Both venues offer a critical perspective on local as well as national and international social, political, and cultural issues, and both tend to prioritize coverage of environmental topics. Thus these two organizations are ideal test cases through which to examine how alternative media engage with climate change.

All items published by these two sources between January 1 and December 31, 2009, that contained one or more of the key phrases *climate change, global warming, greenhouse gas emissions, CO2 emissions, carbon dioxide emissions, carbon tax,* or *Copenhagen* were collected. This search generated 216 pieces from *The Straight* and 145 from *The Tyee*, for a total sample size of 361 items. All items were subject to a qualitative critical discourse analysis that identified dominant themes in the coverage with a particular emphasis upon how climate politics was represented. Key research questions were the following: To what extent was climate change portrayed as a political issue, rather than simply a scientific or technological phenomenon? How did a focus upon specific themes, issues, or frames accentuate (or inhibit) the representation of climate change as political? Who were the primary political actors and sources featured within items and how were they represented? How were political agency and efficacy assigned to particular groups and individuals? How were readers explicitly and implicitly positioned with respect to observing and participating within climate politics? What was the mix between pessimistic and optimistic accounts of climate politics, both within individual items and across the sample as a whole? And finally, what role did normative moral and philosophical arguments play in cultivating more engaged forms of political subjectivity? The focus of critical discourse analysis is the relation between language and power (Fairclough, 2001): In this case, my interest lay in analyzing how alternative media discourse provided the conceptual tools as well as the affective resources to (re)configure climate politics as a means of challenging dominant political and economic structures that currently block real action on climate change.

EXPANDING THE FRAME: POLITICIZING CLIMATE CHANGE

In mainstream news media, attention to the political dimensions of climate change tends to ebb and flow depending upon the presence of key focusing events, such as international summits or, more rarely, the implementation of domestic policies. In the absence of such events, climate politics tends to play a minor, background role in news about other aspects of climate change. Stories about developments in climate science, for example, or the current and future effects of climate change (e.g., extreme weather events) rarely frame these issues in terms of their relevance for climate politics. Coverage of the "green economy" often fails to consider how the deployment of new technologies of energy and transportation might be enhanced as part of an integrated political response to climate change. Profiles of eco-friendly products and advice about how to reduce one's carbon footprint through lifestyle change encourage people to conceptualize their own responses to climate change in a fundamentally individualized and apolitical manner. When climate change does appear as a political issue, it is usually narrowly represented as the domain of national and international political elites,

Visions of Climate Politics in Alternative Media 253

a rarefied space that citizens may passively (and cynically) observe but that lies largely beyond their influence and control. Moreover, the tone of such coverage is often negative and pessimistic, presenting an image of climate politics as irredeemably flawed. For the most part, little positive attention is given to progressive political action, whether in the form of policy-based solutions or that of democratic activism (Gunster, 2011). Therefore climate politics almost always appears as part of the problem and rarely as part of the solution.

In contrast, alternative media offer a much more expansive vision of climate politics that combines extensive investigation into the failures of existing political institutions with compelling and often inspirational accounts of how activist political engagement can transform those institutions into powerful and positive forces for change. During 2009 climate politics was clearly the dominant theme in alternative media coverage of climate change: Most items (over two thirds) had an explicitly political focus; a small number of pieces prioritized either climate science or voluntary, lifestyle-based solutions.

In both *The Georgia Straight* and *The Tyee*, then, climate change was overwhelmingly framed as a *political* issue. Motivating local and national governments to adopt strong climate policies was consistently portrayed as the single most effective—and indeed, the only realistic—means of addressing climate change.

Before digging into the details about how climate politics was framed, it is worth noting that one of the most important contributions of alternative media to expanding the conceptual horizons in which to consider climate change was periodically stepping back from the immediacy of particular policies and events to provide a big-picture view of the broader questions posed by environmental crisis. Alternative media were especially keen to grapple with the deep structural changes that are required in economic and political practices, as well as in society's moral and philosophical values, in order to inaugurate and expand truly sustainable ways of life (Leiserowitz & Fernandez, 2008). Both *The Georgia Straight* and *The Tyee* regularly featured "big ideas" pieces that tried to theorize some of the deeper issues raised by climate change. Many columns and op-eds, for example, presented thoughtful criticisms of the paradigm of economic growth, not only observing the fundamental incompatibility between endless industrial expansion and ecological health but also raising serious doubts about the capacity of this paradigm to serve human needs. In a piece representative of this broader critical focus, University of Victoria environmental studies professor Michael M'Gonigle warned that "success" in the Copenhagen negotiations would further entrench the structures that are responsible for climate change, provocatively describing economic growth as

> the largest elephant in the room because the whole panoply of solutions on the table—cap and trade, carbon taxes, clean development mechanisms, carbon offsets—are all made to fit within capitalism and its growth imperative. But doing without growth is not something anyone is prepared to consider. Growth is the lifeblood of capitalism. We simply dare not, cannot, talk about it.
> (M'Gonigle, 2009c, n.p.)

For M'Gonigle (and others), the only real hope in genuinely confronting the growth fetish lies in a radical democratization of the economic system, dramatically expanding the political sphere in order to bring systems of production and consumption under popular control and regulation. "Actually empowering citizens to try out new things where they live entails a form of what Harvard law professor Roberto Unger calls 'democratic experimentalism'. DemocracyNow! calls it 'deep democracy'" (M'Gonigle, 2009a). Pieces like these that use climate change as a platform to launch wide-ranging conversations about big ideas, such as the need to confront dominant paradigms of economic growth and experiment with new forms of participatory democracy, challenge readers to expand the conceptual horizons through which they understand and engage climate change as a political phenomenon.

Politics (still) matters: Seeking hope in dark times

Given alternative media's commitment to providing alternative and often oppositional perspectives on political issues and events, it comes as little surprise that their pages were filled with strong denunciations of the inaction, obfuscation, and delay perpetrated by governments. The most obvious example of this during 2009 was the uniformly critical coverage of the Canadian government: Of those items that contained an explicit assessment of Canada's performance on climate change, virtually all of them offered an entirely negative appraisal. Among the most persistent critics was scientific icon and weekly *Georgia Straight* columnist David Suzuki, who authored dozens of pieces attacking the government's weak environmental record. What was especially noteworthy about the strong criticism of Canada from Suzuki and others, though, was how often it was framed in *comparative* terms by contrasting domestic weaknesses with the much stronger commitments made and actions taken by others. Criticism of political actors and institutions, in other words, was almost always specific rather than totalizing, and most stories about the failures of one government were balanced with favorable references to the achievements of others.

During the first few months of 2009, the positive foil deployed against Canada was, more often than not, the incoming Obama administration, which was held up by many as evidence of the potential for effective, progressive climate action within existing political institutions. Contrasting Obama's activist actions and intentions with both the reactionary intransigence of his predecessor, as well as the resistance and lethargy of the Canadian government, built upon broader narratives of hope and change that had taken shape around his presidential campaign. The underlying theme of such pieces was that *politics matters*, both in terms of setting (and changing) the policies, procedures, and objectives for a wide range of influential public institutions and regulatory agencies and in terms of shaping the values and priorities for society at large. In May 2009, for example, Suzuki lauded Obama for "giving the American people hope that positive change is possible. If only we were being offered the same kind of hope here in Canada." Detailing these changes, Suzuki explained how the U.S. president "has injected billions of dollars into science ... signed into law protection for over two million acres of wilderness, and made clear his intention to combat climate change, including a willingness to force automakers to produce more fuel-efficient and less-polluting cars" (Suzuki & Moola, 2009d, n.p.). Beyond simply echoing the prevailing fascination with Obama's political charisma, stories like these make the case that electoral politics can and does matter in terms of concretely shaping environmental policy. At one level, this may seem like an uncontroversial and even banal claim. Yet it actually serves as an important counterpoint to an increasingly widespread and cynical pessimism that discounts *all* political institutions as equally ineffective and uninterested in tackling the environmental crisis. It also reminds environmentally minded citizens that electoral campaigns are a critical venue through which to engage in climate politics.

Much has been made of the importance of an emerging "green economy" frame that can challenge the "jobs versus environment" frame that has tended to dominate media coverage of environmental issues over the past several decades (Fletcher, 2009; Zehr, 2009). Environmental communications strategists are especially enamored of the potential for such a frame to bypass divisive (and often ideological) battles over climate change and, instead, cultivate bipartisan consensus on so-called win-win issues, such as the economic and employment opportunities or the increased energy security that will accrue to nations that lead the way in revolutionizing the production and use of energy (e.g., ecoAmerica, Western Strategies, & Lake Research Partners, 2009; "Sell the Sizzle," 2010). Stimulating public awareness and excitement about the possibilities for greening the economy is a crucial task in building public support for action on climate change. However, it is essential that one not neglect the essential role that a hospitable legislative, regulatory, and policy environment plays in facilitating the growth of such activity.

If government is not actively cast as the midwife of this economic and technological transformation, that role will instead be filled by the free market, and glowing reports about the achievements of green enterprises will be taken as evidence of the sufficiency of market-led corporate innovation in addressing the environmental crisis.

Even worse, stories of the green economy may disappear into the business pages, where companies and technologies are assessed first and foremost as investment vehicles, and where public policy (if it is mentioned at all) is measured according to its impact on the bottom line. Opportunities to weave enthusiasm for the green economy into broader narratives about social and economic justice, exploring employment opportunities for low-income and displaced workers that could mobilize diverse political constituencies in support of climate-related policies, are then easily displaced by neoliberal fantasies of wide profit margins and national competitiveness. Whether the green economy is framed as an active political response to economic and ecological crisis or as emerging more or less directly from the entrepreneurial ingenuity of the private sector matters a great deal in terms of expanding or contracting the field for climate politics. Most important, perhaps, the failure or refusal to constellate the green economy with climate change (out of fear of rhetorically "contaminating" the positive associations of the former with the negative associations of the latter) can accentuate the apocalyptic (and antipolitical) framing of climate change as a problem without a solution.

Among the most distinctive characteristics of alternative media coverage I examined were the efforts to foster discussions of energy and climate with a strong emphasis upon government policies, programs, and regulations that framed the green economy as a political matter rather than solely an economic or business phenomenon. Political solutions to climate change (i.e., actions taken by government that are mandatory in nature) attracted considerably more attention than either voluntary and lifestyle solutions or technological solutions. Beyond the strong emphasis upon political solutions per se, a political frame often dominated discussion of other types of solutions: Most discussions of technological solutions, for example, identified their development as contingent upon regulatory, policy, or funding initiatives from government.

In recent years it has become commonplace to attack the environmental movement for its tendency to rely upon apocalyptic rhetoric that prioritizes terrifying accounts of the ecological nightmares humanity faces rather than developing inspirational and inclusive visions of a sustainable future all human beings might share (e.g., Nordhaus & Shellenberger, 2007; Turner, 2007). Although references to the catastrophic potential impacts of climate change were a common feature of alternative media coverage, I found that the overall tone of both news and opinion pieces was generally optimistic and hopeful, balancing the need for urgent and immediate action with the confidence that real, practical solutions do exist. Very few stories were genuinely apocalyptic in the sense of presenting doomsday scenarios of catastrophic global warming as the inescapable fate of humanity. Instead, climate change was consistently described as a problem that lies within humankind's collective power to solve. Especially noteworthy in this regard were those pieces that openly celebrated the utopian dimensions of sustainability actually being practiced today: Filled with positive affect, such stories invited readers to actively participate in imagining how (easily) a better world might be built. Through investigation into the policies and practices of sustainable energy, transportation, agriculture and urban design, *The Georgia Straight* and *The Tyee* provided their readers with a steady supply of broadly optimistic, solutions-oriented journalism.

Thinking globally, acting locally: Rescaling climate politics

According to many scholars of climate change communication, facilitating public engagement with climate change is best achieved at the local level, in terms of both building awareness about the local impacts of climate change and informing citizens about the possibilities for taking effective action at a local level (Leiserowitz, 2005; Lorenzoni & Pidgeon, 2006; Segnit & Ereaut, 2007). In their study of climate change discourse in the United Kingdom, Nat Segnit and Gill Ereaut celebrated the capacity of local frames "to make climate change, and solutions to it, real, *tangible*, in a way that is clearly far harder at the national and/or international level" (2007, p. 32). Writing more broadly about all forms of environmental politics, Christopher Rootes noted that "acting locally will not usually be enough to secure redress of environmental grievances but, for most people, local campaigns offer the only accessible entry to the political struggle for ecological sustainability" (2007, p. 739). Yet as Carvalho pointed out, "sustained analysis of the possibilities for local policy making on climate change features only rarely in mainstream media" (2010, p. 76). Instead, climate politics is usually represented as something that occurs behind closed doors in national and international forums, far removed from the influence of ordinary citizens.

The Georgia Straight and *The Tyee*, however, challenged this narrow framing of climate politics by exploring the multiple scales on which climate politics occurs, paying particular attention to the climate-related policies of the provincial government. Unlike coverage of federal climate policy and politics that was heavily clustered around Canada's disappointing performance during the Copenhagen summit, focus upon provincial policies and politics was more consistent through the year; alternative media featured a wide range of news and opinion analyzing the actions of the British Columbia government from a variety of perspectives. Roughly one third of items *directly* engaged the question how well or poorly the provincial government was performing on the issue, thereby reinforcing the connection between climate change as a global environmental crisis and local political institutions as *the* most relevant sphere of action for those interested in the topic. Assessment of the British Columbia government's performance was more balanced than the entirely negative reviews of the federal government given the more progressive approach to climate policy taken by British Columbia in some areas, especially its introduction of a carbon tax in 2008. Yet the majority of coverage adopted an overall perspective that was critical of the provincial government.

The overarching theme through most of the coverage of provincial action on climate change was the rather stark contradiction between seemingly aggressive climate policies on the one hand and a variety of other initiatives resulting in increased provincial emissions on the other hand. As many pieces noted, the provincial Liberals deserved credit for actions such as introducing a carbon tax, legislating deep cuts in emissions by 2020 and 2050, joining the Western Climate Initiative, and requiring all public-sector operations to achieve carbon neutrality (through reductions and offsets) by 2010. At the same time, however, the province was also continuing to provide massive subsidies to the local oil and gas industry, spend billions on expanding highways as part of the Gateway infrastructure program, and quietly supporting the construction of the Enbridge pipeline across northern British Columbia to enable the direct export of oil from the tar sands to Asian markets. Inclusion of perspectives from environmental activists, social justice advocates, and opposition politicians, both in the form of op-eds and as sources within news items, provided the public with an ongoing account of provincial policies that was both detailed and holistic in nature, ensuring that the Liberal spotlighting of particular initiatives, such as the carbon tax (for which Gordon Campbell received an award in Copenhagen), did not displace attention from how other provincial policies were in fact contributing to increased emissions. In addition to leaving citizens better informed about the overall effect of provincial policies, this holistic approach can cultivate a more sophisticated understanding that achieving real reductions in emission levels requires coordinated action across a broad variety of areas. Ultimately, though, the most important effect of this coverage may have simply been to hammer home the message—again and again—that the policies of the provincial government have an enormous impact upon local emission levels and that, accordingly, participation in provincial politics (either within the conventional political process through provincial political parties and election campaigns, or outside that process through activism targeted at the provincial government) provides an excellent opportunity for citizens to directly participate in climate politics.

The Gateway infrastructure program received considerable attention during 2009. As noted, it served as a lightning rod for those criticizing provincial government spin for exaggerating the impact of a few key policies, such as the carbon tax, and also minimized the ways other government actions would contribute to increasing emissions levels. Many pieces focused predominantly on the negative aspects of the project, including the role of highway expansion in facilitating low-density, high-emission patterns of suburban growth and its destructive impact upon wildlife habitats. The most compelling criticisms of Gateway, however, were those that (re)framed the project as a *political choice* between increasing reliance upon existing forms of automotive transport (the largest source of emissions in British Columbia) or switching to new systems of mass transit. In March, *Tyee* editor David Beers wrote a lengthy article discussing a study by Patrick Condon, a professor with the UBC Design Centre for Sustainability, that found that the $3.1 billion allocated to a single new 10-lane bridge (to replace a smaller 5-lane structure that still has 30 years remaining in its lifespan) could, instead, be used

> to finance a 200-kilometre light rail network that would place a modern, European-style tram within a 10-minute walk for 80 per cent of all residents in Surrey, White Rock, Langley and the Scott Road district of Delta, while providing a rail connection from Surrey to the new Evergreen Line connecting Pitt Meadows and Maple Ridge into the regional rail system
> (Beers, 2009b, n.p.)

Rather than simply presenting a list of complaints about the project and demanding that it be stopped, this article offered a positive alternative vision of how public resources could be more efficiently and effectively allocated. It challenged members of the public to "think big" in terms of how they might make neighborhoods and communities more livable and environmentally friendly, kick-starting a process of imagining what a more sustainable future might actually look like. Equally significant, drawing attention to the enormous cost of maintaining and expanding existing structures of automobility recasts such (utopian) visions of sustainability as eminently reasonable and practical, well within the power of governments to achieve. Most important, technocratic decisions about infrastructure are repositioned as political choices that ought to involve a democratic discussion, debate, and choice among different options (and the broader values they invoke). "A real comparison between reasonable alternatives," explained Condon, "would enhance our ability to choose wisely. How we deploy the available billions on transportation infrastructure over the next 10 to 15 years will determine how sustainable this region will or won't be 100 years from now" (cited in Beers, 2009b, n.p.). Beyond this initial article, the bridge versus light rail network meme was reproduced in a number of other news and opinion pieces (e.g., Dembicki, 2009; Kimmett, 2009a; Kettlewell, 2009; M'Gonigle & Anderson, 2009) that consequently helped recapture the conceptual high ground of "progress" from those defending the project: Instead of appearing simply as naysayers trying to stop development, Gateway's opponents could represent themselves as far-sighted proponents of truly sustainable development based upon practical and cost- and carbon-effective public transit. In contrast, *The Vancouver Sun*, the province's leading daily broadsheet, did not include a single reference to the Condon study throughout the entire year.

"The real solutions" to climate change, argued Rex Wyler,

> will demand a paradigm shift as dramatic as when Copernicus pointed out that the universe did not revolve around earth. The answer will be in localization, based less on foreign-made goods, debt, and commuting, and more on friends, local food and commu-

nity cohesion. The new normal will be about improving the quality of life without consuming more stuff.

(Wyler, 2009, n.p.)

Optimistic paeans to localization were common in alternative media, especially in the pages of *The Georgia Straight*, which frequently championed the idea of downscaling to a local level as the foundation upon which sustainable forms of living might be constructed. This theme dovetailed neatly with the program and priorities of the provincial Green Party: Whereas the policies of the Greens were largely ignored by mainstream media, they received considerable attention from *The Georgia Straight* during the May provincial election (e.g., Hui, 2009; Burrows, 2009a; Lupick, 2009), including an endorsement from the paper's editorial board (*Straight* editorial staff, 2009). In a guest op-ed on the eve of the election, Damian Kettlewell, the party's deputy leader, argued that

> the greatest job opportunities for the current and next generation of British Columbians will come from localized markets that are based less on foreign-made goods, less on assuming large debt, and less about commuting. These locally-based jobs will be increasingly based upon local communities, local food, and the genuine needs of local people.
>
> (Kettlewell, 2009, n.p.)

Such sentiments were taken up in a more substantive and engaged fashion by *Straight* editor Charlie Smith (2009) in a lengthy feature news article—at over 2,500 words, it was one of the longest items to appear in alternative media in 2009—that explored how "a burgeoning relocalization movement has the potential to revolutionize the way we eat, shop, work, and vacation" (n.p.). The piece brought together a wide range of examples of increasingly localized consumption practices from the food, clothing, entertainment, building, and manufacturing sectors with a more philosophical discussion inspired by recent bestsellers critiquing globalization, including Jeff Rubin's *Why Your World Is About to Get a Whole Lot Smaller: Oil and the End of Globalization* (2009), John Ralston

Saul's *The Collapse of Globalism and the Reinvention of the World* (2006), and Naomi Klein's *The Shock Doctrine: The Rise of Disaster Capitalism* (2007). An excellent example of the "big ideas" genre discussed earlier, the item developed and applied the concept of localization at a variety of levels: first, as a means of describing empirical trends in the production and consumption of local goods and services; second, as an increasingly important normative concept through which individuals were coordinating lifestyle, identity, and political values; and third, as an economic and political ideal, connected with related ideals of democracy, social justice, and sustainability that could be realized through a range of policies and regulatory structures at the local, national, and international levels.

From cynical observer to engaged citizen: Normalizing climate activism

The most striking difference between alternative and mainstream media's presentation of climate politics was the extensive and largely sympathetic coverage that the former provided to climate activism. The single most consistent theme running through virtually all coverage of climate change was the need for more political will to address the problem. As one young Greenpeace activist put it,

> The voices of this generation are screaming for vision, leadership, and change ... The environmental movement has identified solutions. It is urging governments to move into a clean, green energy economy that will bolster the economy, create jobs and improve our environment ... We have the scientific analysis. We have the green technology. We have the desire and the need. What we must have now is the political will. For the sake of this generation and the ones to come, we must find it.
>
> (Wilson, 2009, n.p.)

Suzuki often ended his columns with similar calls to action: "If we speak loudly enough, they will listen" (Suzuki & Moola, 2009a, n.p.);

"we can all take individual action to reduce our emissions, but ultimately we must let our leaders know that we expect them to seize the opportunity in Copenhagen" (Suzuki & Moola, 2009d, n.p.); "visionary leadership requires active and engaged citizens to keep the politicians' feet to the fire. Your efforts have never been needed more to make this happen" (Suzuki & Moola, 2009b, n.p.). Whereas mainstream media often deploy a "regime of objectivity" (Hackett & Zhao, 1998) that positions the audience as passive consumers of news, alternative media both assume and interpellate a readership whose members are eager to engage climate politics as active citizens. Whereas mainstream media often depict climate politics as a reified sphere in which the capacity to exercise political choice has been stripped from individuals, alternative media adopt a moralized language of crisis, agency, and collective responsibility, insisting that the choice to mobilize global resources to confront climate change or to continue with business as usual is one that properly belongs to everyone. Instead of presenting neutral, impartial accounts of the world, alternative media are far more likely to use facts and figures as the building blocks of argument and opinion, normalizing an active and engaged orientation to the world in which one is motivated and even compelled to make political choices and assume some measure of political accountability for them. This is not to say that polemic entirely displaces journalism but, instead, that the luxury of being a casual, disinterested observer of climate politics is a much harder identity to sustain in the discursive universe of alternative media.

At times, the demand for increased political engagement appeared more rhetorical than substantive, with calls for political activism coming across as somewhat generic and formulaic rather than substantive. This was especially the case when such demands were not accompanied by any concrete advice or even speculative consideration about how citizens might actually participate. Urging the public to get involved in climate politics is one thing, but explaining the different ways they actually can get involved is something quite different. As Tzeporah Berman put it in an interview with *The Georgia Straight*, "We now have an over-developed consumption muscle and an under-developed civil-engagement muscle" (Burrows, 2009b, n.p.). Passionate but hollow affirmations of the virtues of collective political action risk becoming little more than activist clichés, rhetorical fetishes that may intensify the public's craving for political agency but leave them no better informed about how they might act to satisfy it. Indeed, moralized injunctions to "get active" that fail to teach people how to make the transition from passive observer to engaged citizen can, paradoxically, intensify perceptions of political alienation and guilt, leaving individuals feeling even more apathetic, dispirited, and disengaged.

Fortunately, however, the rhetorical celebrations of climate activism that often appeared in alternative media were backed up with detailed, specific, and even occasionally narratively grounded accounts of activist political engagement. References to collective political climate activism, including demonstrations, sit-ins, petitions, letter-writing campaigns, and so on, appeared in one of every seven stories on climate change. That these stories on climate change featured *actual practices* (rather than simply abstract notions) of citizen engagement suggests the key role that an activist orientation played in alternative media representations of climate politics. Most important, it transformed such politics from a spectacle to be observed into a *site of struggle*, a fluid and dynamic space of political action in which members of the public can and do intervene to shape the policies and priorities of government. This emphasis stands in marked contrast to mainstream media's tendency to ignore and marginalize attention to political activism. Research on British Columbia regional media coverage of the Copenhagen summit confirms the significant differences between the two types of media in this regard: Whereas collective political actions were mentioned in only 6% of the items from the two major daily newspapers, they appeared in close to 35% of alternative media stories during the same three-week period (Gunster, 2011).

The steady accumulation of stories about climate activism can grow into powerful narratives of collective mobilization and political momentum in which particular examples of climate activism are framed as both expressive and constitutive of broader trends in public opinion, as well as in transformative social movements. Immediately after describing recent polls showing high public support for stronger emission targets, *Tyee* reporter Colleen Kimmett explained that

> there have been arrests on Parliament Hill, occupations of MP's offices and protests around the country. People who have never waved a banner in their lives are joining in the protest of how Canada is behaving in Copenhagen, including a minister in Toronto who has sworn to fast for seven days.
> (Kimmett, 2009b, n.p.)

The shape of public opinion on climate change, in other words, was directly linked to the actions of protesters, reversing the polarity that otherwise often accompanies the representation of protest as qualitatively different from the views of most people. In a similar vein, Suzuki noted,

> it's been heartening to see so many people, especially young people, taking to the streets and Parliament Hill, writing to MPs and prime ministers, and joining campaigns to urge governments to be a part of the solution to global warming. Millions of people turned out recently for more than 5,000 International Day of Climate Action events in 180 countries.
> (Suzuki & Moola, 2009c, n.p.)

Bringing a broadly global and historical lens to the political activism on display in Copenhagen, Naomi Klein cast the emerging movement for climate justice as challenging the power of corporate capitalism by offering a compelling vision for how societies could be organized differently:

> activists in Copenhagen won't simply say no to [corporate capitalism]. They will aggressively advance solutions that simultaneously reduce emissions and narrow inequality. Unlike at

previous summits where alternatives seemed like an afterthought, in Copenhagen the alternatives will take center stage.

(Klein, 2009, n.p.)

Analyses like this perform the vital function of knitting individual actions together into a political tapestry that enables those involved in such actions, as well as those sympathetic to them, to understand the long-term significance of those actions as going far beyond their specific or immediate impacts in terms of the contribution to building a popular movement for climate justice.

Moreover, Klein's astute expansion of the activist frame to include alternatives and solutions as well as protest and anger is reflective of the more complex portrait that alternative media provided of climate activism. Political activism is not characterized simply as protest but also as a means of engaging in a constructive, grassroots dialogue about climate solutions. Deliberative forums, such as the People's Climate Summit (*Klimaforum 09*)—a gathering of hundreds of nongovernmental organizations (NGOs) from around the world, including environmental groups, labor unions, farmers, students and local community organizations—received considerable attention in *The Tyee* and *The Georgia Straight*, illuminating a positive, solutions-focused dimension to climate activism (e.g., Beresford, 2009; Hiskes, 2009; Ravensbergen, 2009). In direct contrast to the gridlock and intransigence of the formal negotiations, these stories helped (re)establish the viability of setting priorities, building policies, and negotiating compromises through creative and innovative processes of consultation, dialogue, participation, and collaboration, each of which was guided by values of empathy and the common good rather than by personal and national self-interest. Challenging representations of the primary divisions of climate politics as only constituted through national identity, these forums (and alternative media coverage of them) offered alternative political narratives defined by emerging affinities between "ordinary" people from all countries based on basic principles of justice and equity. Beyond the virtues of any particular proposal or policy, the real significance of such coverage lay in how it challenged dominant images of climate politics as a space of delay, division, and failure, instead providing compelling evidence that alternative forms of political engagement are not only possible but actively practiced by those committed to charting a different course on climate change.

Although environmental activists are often attacked as harbingers of doom, the infusion of activist sentiments into the environmental journalism of alternative media helped produce news and opinion that was, on the whole, far more hopeful than that which tends to originate from mainstream media on this issue. *The Georgia Straight* and *The Tyee* opened their op-ed pages to political activists directly engaged in the politics of climate change at both a global and a local level. Unlike mainstream media, in which activists typically appear (if at all) as one source or sound bite among many, alternative media provided them with ample space to weave their ideas and their passion into a coherent set of arguments and—more important, perhaps—allowed them to serve as powerful exemplars of an engaged political subjectivity. I end this section with a closer look at one piece from a local activist that brilliantly fused a critical analysis of climate politics with optimistic accounts of grassroots climate activism as a growing social force.

As the Copenhagen summit was drawing to a close, *The Tyee* published an op-ed by Jamie Biggar (2009), who was in Denmark as a member of the Canadian Youth Delegation, a group of young Canadians who traveled to Copenhagen to pressure the federal government to play a more constructive role in the negotiations. Biggar's piece was published in the form of a dialogue with university professor Michael M'Gonigle who, as noted earlier, had argued (2009a, 2009c) that it would be better for the Copenhagen negotiations to fail than produce an agreement that left dominant economic and political structures fundamentally untouched. Real progress on climate change, M'Gonigle asserted, was impossible until core questions of economic growth, and the role of states and markets in protecting and promoting it, were addressed. Though Biggar largely agreed with these claims, he also argued that climate politics is increasingly filled with examples of activism and political radicalization, which make such transformations more rather than less likely. "Climate treaty summits have a funny way of radicalizing people whose gut instinct is to be part of the establishment," he wrote, citing climate scientist James Hansen's adoption of the tactics of civil disobedience to challenge the U.S. coal industry. "The list of establishment types who no longer believe that regular channels can solve the problem is getting longer" (Biggar, 2009, n.p.) After a thoughtful and extensive reflection on the question "How do we get what we can from the dominant institutions that exist today while we build the dominant institutions that we need tomorrow," Biggar ended with a hopeful reflection upon his own experience as an activist. "I'm thinking about the thousands of Canadians," he mused,

> that have worked together over the last months and are now making phone calls and hitting the streets—many of whom, maybe most, had never been politically engaged before in their lives. I know that for many of these people, certainly for myself, this is the first time that they have felt like they were part of something so much bigger than themselves. There's something powerful brewing, M.M., something that almost shakes with its energy and potential.
>
> (Biggar, 2009, n.p.)

Brilliantly shifting between a normative injunction to practice a different kind of politics and an empirical description of those politics in action, pieces such as these invite the public to join like-minded others in a collective, political response to the failures and shortcomings of existing political institutions. Activism, not apathy, was again normalized as the most logical and the most satisfying response to a pervasive crisis of political alienation.

In a fascinating study of the impact of social norms upon environmental behavior, Griskevicius, Cialdini, and Goldstein (2008) distinguished between injunctive norms (prescriptions of how people *should* behave) and descriptive norms (descriptions of how people *are* behaving), arguing that the latter are far more effective than the former in motivating and shaping behavior. Showing people that others—preferably, others *like them*, with whom they can identify and empathize—are engaging in a particular form of action is a far better means of persuasion than simply explaining or asserting the need for or the benefits of that form of action. Indeed, when such explanations are framed in negative terms (as they often are)—for example, it is terrible that the Canadian public is not more involved in climate politics because it is an issue that affects everyone—they can have the perverse effect of strengthening the undesirable behavior because it is represented as normal. Griskevicius applied these insights to everyday behavior, but they are equally helpful in thinking about how to expand the forms of political action that citizens consider normal. According to this logic, the best means of increasing civic engagement would be to represent such behavior as common, widespread, pleasurable, and politically effective: in short, as *normal*. Stories are needed that feature testimonials from individual activists describing not only their moral or political convictions but also, and more important, narrative accounts of their personal experience of becoming politically active and how that experience has made them feel, has affected their identity, and has changed how they understand and engage with the world. Such narratives help close the gap between citizens and activists, smoothing the transition from the former to the latter as people come to see different forms of democratic engagement as normal activities that others just like them are doing (and enjoying) in order to express and act upon their political beliefs and moral values.

FACING THE FUTURE: ENVIRONMENTAL APOCALYPSE, CLIMATE POLITICS, AND ALTERNATIVE MEDIA

Some might fault the vision of alternative media I have developed here as both selective and idealized, an overly optimistic portrait that exaggerates the potential of this medium to lead the way in motivating civic engagement with climate politics. And there is probably some truth to that criticism. Rather than provide a systematic overview of how alternative media represent climate change, my goal has been to explore the most praiseworthy features in its coverage of climate politics. In particular, I wanted to showcase its tendency to challenge the sclerotic, hidebound, and thoroughly cynical visions of such politics that so often dominate the public sphere by reminding the public that, to invoke the slogan of global social justice, another world *is* possible. Governments *can* implement more active climate policies, not simply because they should, but because other governments *are* implementing (and benefitting) from such policies already; more important, citizens *can* become (more) active in climate politics not simply because they should, but because thousands—millions—of others like them *are* active in countless different ways. Prescription and description come together in a virtuous iterative cycle in which alternative media use stories of both policy and activist success to fuel ever increasing levels of civic engagement.

My emphasis upon success stories should not, however, be understood as endorsing the increasingly common view that climate change communication must abandon "alarming" talk of rising sea levels, global drought, species extinction, and climate chaos for sunnier, bipartisan narratives about energy independence, green jobs, and sustainable communities. Indeed, much of the rhetorical force of alternative media discourse on climate change arose precisely from its deliberate juxtaposition of those two very different visions of humanity's collective future. After all, heaven and hell each depend upon the other for their meaning and significance. As Futerra Communications, a British firm specializing in climate change messaging, noted, "Today we have a choice between that positive picture [of sustainable societies] and the alternative of unmitigated climate change. It's extremely important to hammer home that this moment is the moment of choice between the two paths" ("Sell the Sizzle," 2009, p. 5)

Visions of Climate Politics in Alternative Media 273

Alternative media never pulled punches when it came to referring to the impacts of global warming, upon either British Columbia or the world more broadly. Columnist Gwynne Dyer, for example, soberly described a possible rise in average global temperatures beyond two degrees Celsius as

> the point of no return. Further warming would trigger natural processes that release vast quantities of greenhouse gases into the atmosphere from melting permafrost and warming oceans. These processes, once begun, are unstoppable, and could make the planet four, five, or six degrees hotter than the present by the end of the century. At those temperatures, much of the planet turns to desert, and the remaining farmland, mostly in the high latitudes, can support at best ten or 20 percent of the world's current population.
>
> (Dyer, 2009a, n.p.)

As the Copenhagen talks collapsed, Dyer explained that "each year in which we don't reach an adequate global climate deal is probably costing us on the order of fifty million premature deaths between now and the end of the century" (2009b, n.p.). Frequent references by Dyer and others to global warming "tipping points"—thresholds that, once crossed, activate positive feedback loops that take away human ability to do anything to mitigate catastrophic climate change—were extremely effective in making the case for immediate action. Irrespective of how dramatically they may be presented, linear accounts of worsening climate impacts (e.g., rising sea levels, declining access to fresh water) always leave open the possibility of taking action at some point in the future. In contrast, the prospect that today's inaction may *permanently* close that window of opportunity and condemn this generation (and the next) to a terrifying world of climate chaos demands that people either take action now or forever lose the chance.

Drawing upon scholarship in literary and rhetorical criticism, Christina Foust and William Murphy (2009) distinguished between tragic and comic forms of apocalyptic climate change discourse, which they argued represent agency, temporality, and telos in qualitatively different ways. Viewing apocalypse tragically suggests that human agency is limited to "following the divine will and behaving in ways decreed by God" (Wojcik, 1996, p. 314), moving toward a catastrophic telos that is clear and unstoppable. Taking a comic perspective, humans are responsible for a course of action, giving them some play in influencing their fate (though not totally changing the disastrous outcome foretold, an outcome which is more ambiguous than the tragic telos; p. 154).

The realities of climate science leave little room for honest discourse about climate change to be anything other than broadly apocalyptic in nature: there is simply no other way to truthfully represent the situation humankind now finds itself facing (Intergovernmental Panel on Climate Change, 2007b; Risbey, 2008). However, whether the apocalyptic character of such discourse is tragic or comic may depend much less on how climate impacts are portrayed and much more on how the *politics* of climate change is represented. At their best, alternative media position civic engagement and collective action as a chance to avoid disaster. And therefore alternative media constitute an invaluable ally and resource for climate change communicators in motivating and mobilizing deeper, broader public awareness of climate politics, as well as the many ways in which people can get involved in the fight against climate change.

ENDNOTES

1. This chapter is part of the Climate Justice Project, a SSHRC-CURA research study led by the Canadian Centre of Policy Alternatives—British Columbia and the University of British Columbia. Funding for this research was provided by the Social Sciences and Humanities Research Council of Canada; the Office of the Vice President, Research at Simon Fraser University; and the Office of the Dean, Faculty of Communication, Art, and Technology. The author would like to thank Shannon Daub, communications director at CCPA-BC, for her advice on study design.
2. For a rare exception to this trend, see Antonio López's (2010) argument that local, organic media are much better situated than mass, industrial media to cultivate political engagement with climate change.

CHAPTER 10

POVERTY, PROTEST, AND POPULAR EDUCATION

CLASS INTERESTS IN DISCOURSES OF CLIMATE CHANGE

Eurig Scandrett, Jim Crowther, and Callum McGregor

All environmental-ecological arguments … are arguments about society and, therefore, complex refractions of all sorts of struggles being waged in other realms.… Each one of these composite discourses shapes a unique blend of complicity and dissent with respect to existing beliefs, institutions, material social practices, social relations, and dominant systems of organising political-economic power. This is their specific virtue: they pose problems of defining relations across different moments in the social process and reveal much about the pattern of social conflict in all realms of social action.

(Harvey, 1996, p. 373)

Introduction

The phenomenon of accelerated anthropogenic greenhouse gas emissions leading to increased average temperatures on the earth's surface was identified in the 1970s and recognized internationally as an environmental problem by the World Meteorological Organization at its World Climate Conference in 1979. Climate scientists, insofar as they can be considered a homogenous epistemic community,[1] remain at the center of climate change discourses, not least through the powerful position of the Intergovernmental Panel on Climate Change (IPCC) established in 1988 (Simms, 2005; van Beukering & Vellinga, 1996).

However, international policy makers and a growing number of environmental nongovernmental organizations (ENGOs) took up climate change as a key issue. The Club of Rome published its seminal *Limits to Growth* in 1972; the World Wildlife Fund (WWF) was founded in 1961, Friends of the Earth in 1969, and Greenpeace in 1972. The ENGO sector had emerged as a significant social movement (at least in North America and Europe) by the time climate change was recognized, and this issue became an important component of its campaigning program. As these organizations became institutionally established, looser networks of direct action (DA) environmentalists, who focused substantially on climate change, emerged (Seel, Paterson, & Doherty, 2000).

The year 1972 was also the date of the United Nations (UN) Conference on the Human Environment in Stockholm, now widely regarded as the point when environmental issues became a recognized area of international social policy (Jamison, 2001; McCormick, 1991). The UN Environment Program was established following the Stockholm conference and was a key recipient and subsequent generator of policy-led research and debates on climate change in the early 1980s, when policy entrepreneurs in the scientific community were mobilizing on the issue.

It is important to recognize the origins and pathways of climate change thinking because knowledge always carries with it accretions from the social groups through whose hands and heads it passes. Moreover, this process serves social interests through what is assumed, what is ignored, and who is positioned in the discourse to speak (Hannigan, 2006; Harvey, 1999; Pettenger, 2007). The sectors with the most influence on the development of climate change discourses are the scientific epistemic community, the environmental movement, and the UN international policy community (Demeritt, 2009; Giddens, 2009; Hannigan, 2006). Moreover, although these sectors represent discrete sets of interests in themselves, they are also disproportionately drawn from similar social classes: the professional class of educated, salaried workers with technical, managerial, or service functions (the professional-managerial class identified by Ehrenreich & Ehrenreich, 1977; see discussion in Burris, 2005). In the period between the Second World War and the neoliberal turn that began in the 1970s, this class occupied a powerful position in global class dynamics.

Other social class interests contribute to climate change discourse. National and transnational capitalists have mobilized around what is alternately constructed as a threat or an opportunity (see Paterson & Stripple, 2007, p. 151) in order to promote their interests in capital accumulation and political power. The trope of sustainable development has been useful for this class realignment in favor of the "transnational capitalist class" (Sklair, 2001), especially in the decade 1992–2002 between the UN Conference on Environment and Development in Rio de Janeiro and the World Summit on Sustainable Development in Johannesburg, during which time UN policy changed markedly from state-led stakeholder involvement to corporation-led entrepreneurial voluntarism.

In this context, sustainable development emerges as an uncontentious but politically vacuous solution to tackling climate change as a global "postpolitical problem" for which "we" are all responsible (North, 2010; Paterson & Stripple, 2007; Swyngedouw, 1997). As a consequence, the heterogeneous experiences of various subaltern classes are marginalized (H. A. Smith, 2007; W. D. Smith, 2007). Although NGOs have successfully placed "climate justice" or "climate equity" on the negotiating agenda, national and international working class interests, the urban and rural poor, peasants, and indigenous communities have barely influenced the language, values, and analyses that dominate policy and political discussion of climate change.

This generates an issue of "cognitive justice" (Santos, Sousa, Nunes, & Meneses, 2007, p. xxix), which is to say that a "seat at the table" (let alone the driving seat) for subaltern voices in the production of climate change discourse requires that the "logic of the monoculture of scientific knowledge and rigor ... be confronted with the identification of other knowledges and criteria of rigor that operate credibly in other social practices regarded as subaltern" (p. xxix). In a context where statistical abstractions such as the climate, as understood by the IPCC, construct scalar understandings utterly reliant upon complex computer-modeling processes (Demeritt, 2001, p. 309), it becomes difficult to develop bottom-up participation with these specific constructions of knowledge.

The discourse of climate justice is ostensibly well developed and integrated into the UN discussions, although it is not dominant and remains subsumed within a market framework that reinforces inequalities between rich and poor through carbon trading (Lohmann, 2008). Indeed, the incorporation of such narratives into a "new discursive compromise" of "reformist civic environmentalism," along with its invited spaces of participation through multistakeholdership (Bäckstrand & Lövbrand, 2007, p. 134), constitutes an attempt to deliver concessions to subaltern interests without threatening vested power bases. The presence of large summits paralleling the official ones (for example, Cancún in 2010) is indicative of the voice of subaltern groups seeking to present alternative accounts based on their experiences—"invented spaces" as opposed to "invited spaces" for participating in the discourse (Miraftab, 2004). The resultant "People's Agreement" produced an understanding of climate justice that (among many other more particular claims) directly opposes the commodification of knowledge and intellectual property rights that restrict knowledge transfer, does not abandon the notion of recovering and learning traditional wisdom and living in harmony with "Mother Earth," demands the foregrounding of a gender analysis, rejects carbon markets and any role for the World Bank in climate change governance, and addresses "excessive production and consumption patterns" (Friends of the Earth International, 2010, n.p.). Understanding climate change in the absence of significant participation by subaltern classes is a problem principally for three reasons.

First, policy that attempts to tackle climate change and excludes the vast majority of the poor will not be effective. Policy initiatives are just one component of social processes that, if they exclude subaltern interests, will likely produce class conflicts in other areas. There will be resistance, if only through the "weapons of the weak" (Scott, 1985), by which the oppressed seek to obtain marginal benefits with minimal risk through neglect, noncompliance, or even sabotage. There are also potential divisions among subaltern interests, which will be exploited by those who benefit from the ways policy is framed.

Second, climate change already affects the poor to a greater extent than it does the rich. In order to understand how climate change is socially constructed, one needs to start with those who suffer most from its effects. Moreover, social movements that emerge from ecological distribution conflicts provide a distinctive critique by opposing the logic of economic development with incommensurable languages of valuation (Martinez-Alier, 2002).

Third, in order to achieve a socially just solution to climate change, a transformation of economic purpose is needed. Because capitalist accumulation is a core generator of climate change and a producer of poverty and exploitation, a socially just and ecologically sustainable solution needs to prioritize the interests of the subaltern classes. There can be no effective solution to climate change unless the interests and experience of the subaltern classes are included.

Our analysis situates climate change thinking in terms of discourses that are in dialectical relationships with structural forces—especially of class—and in which power, values, beliefs, institutional logics, cultures, and material practices are internalized (see Harvey, 1996, p. 80). Although discourse is an important factor in political struggle (Laclau & Mouffe, 2001; Mouffe, 2005) it is not the only significant one. Nonetheless, as educators, we are especially interested in the function of discourse in the generation, distribution, and production of knowledge and culture. As popular educators we are also interested in the *pedagogy of the oppressed* (Freire, 1972)—that is, forms of education and learning in communities in which "knowledge from below," in critical dialogue with other sources of knowledge, is an integral part of struggles for social justice. Our focus therefore is on how discourses play out in local areas of contention and in community struggles in the context of national policy—rather than in international policy negotiations.

We further analyze discourses of climate change through identifying the diverse narratives they comprise and relate these to major ideological formations with class agendas: social democracy, neoliberalism, ecological modernization, socialist transformation. These somewhat mirror the discourse categorization of Bäckstrand and Lövbrand (2007), but our terminology emphasizes the material basis of interests within the discourse. This helps us explore barriers and opportunities between climate change narratives and the "thematic universe" of subaltern groups. Three Scottish cases are described and discussed that involve dialogues between the narratives of environmental activists and communities living with poverty and the consequences of climate change: (1) the environmental-justice campaign of Friends of the Earth Scotland (FoES), (2) a community project funded by the Scottish policy-led Climate Challenge Fund, and (3) the interactions between DA environmentalists and working-class communities. Finally, we assess the proposal that popular education is valuable for furthering a climate change discourse that includes the interests of subaltern classes, and we argue for a wider employment of this dialogical approach among activists and policy makers addressing climate change.[2] Our account combines reflections from 15 years of observation and participation in environmental justice campaigning by Eurig Scandrett, research into learning in environmental-justice campaigns led by Jim Crowther, and research conducted toward a doctoral thesis by Callum McGregor.

NARRATIVES OF CLIMATE CHANGE

Narratives are the various story lines, shared meanings, and collective assumptions employed in thinking about climate change. In contrast, *discourses* are generally coherent, metalevel frameworks that explain and encompass a wide variety of values and beliefs that draw on amalgams of narratives. The meanings of narratives vary in ideological character according to how they are deployed in discourses, whereas discourses are aligned with the interests of particular social groups. However, both narratives and discourses are processes, never fixed outcomes, and influence one another and other moments of social change (Davis, 2002; Fairclough, 2006; Hannigan, 2006).

It is possible to identify a range of narratives within climate change communication and explore the matrix of interests with which they are associated, thereby indicating whose interests are reinforced through these narratives and whose are ignored. The analysis can be used to identify and differentiate discourses of complicity and of dissent, both between and within these narratives (see Harvey, 1996), and thereby suggest opportunities for intervention.

Generally associated with a liberal emphasis on individual action and personal behavior, *lifestyle choice* is a major narrative of climate change that varies from a minimalist, good-citizenship approach that encourages citizens to "do a little, change a lot" (Scottish Government, 2004) to consumer versions in which products are marketed as climate friendly—and on to more demanding calls for individual carbon rationing. Across these narratives of lifestyle choice there is a common focus on individuals' making a difference themselves. Likewise, a wide range of *technological* narratives is used in end-of-pipe thinking, as in carbon capture and storage or the more demanding power-down accounts of the Zero Carbon Britain project (2010), as well as through laissez-faire approaches to sustainability, in which green technology is expected to be driven by market forces (climate change as opportunity) to generate less damaging solutions. At the far end of laissez-faire is *climate change denial* (climate change as threat), whose advocates generate narratives that obfuscate (Lawson, 2008), or make pseudoscientific assertions about the nonproblematic nature of climate change as a priority in relation to other areas of human development (Lomborg, 2001), or are even militant particularists (e.g., "fuel protestors"). McCright and Dunlap (2000, p. 518) identified three consistent features of denial discourse: a questioning of the strength of the evidentiary basis, a foregrounding of the upsides of climate change, and an assertion that the enactment of climate change policy would do more harm than good. Moreover, in a later article (Dunlap & McCright, 2010, p. 250), they noted that denial narratives are driven by the ideology of a conservative movement operating at some remove from the economic interests that are often used to legitimate their claims.

Narratives of *austerity* are also used in individualized approaches to carbon rationing and reduction, anticonsumerism, and the pietistic approaches to "simple living" (e.g., Sandlin & Walther, 2009), as well as in the more utopian alternative communities. Narratives of austerity frequently make use of conceptual metaphors related to finance ("carbon budgets"), religion ("carbon sinners"), and diet ("low-carbon diets"; Nerlich & Koteyko, 2009). Such narratives can range from the UK government–sponsored Sustainable Development Commission's *Prosperity without Growth* (Jackson, 2009) to the Camp for Climate Action (CCA), which has historically struggled to balance social justice agendas with antigrowth, anticapitalist, and anarchist analyses opposed to the extraction of any fossil fuels (Saunders & Price, 2009; Schlembach, 2011). *Communitarian* approaches vary from ecovillage proponents to transition towns. Transition in this context constitutes a 12-stage process of mobilizing communities to take action to reduce oil dependency (as resilience to "peak oil" and climate change prevention), starting from "easy win" community projects and lifestyle choices and leading up to a more utopian "energy descent plan" (see Hopkins 2008, 2011). Some radical utopians also add an *apocalyptic* narrative, which ranges from DA climate campers to the more resigned Dark Mountain Project (Kingsnorth, 2009). Arguably, this framing has declined because apocalyptic narratives were, for a period, heavily deployed for attention grabbing in popular-media coverage of climate change (Bird, Boykoff, Goodman, Monbiot, & Littler, 2009; Foust & Murphy, 2009; Segnit & Ereaut, 2007). However, elements of the DA movement incorporate a strong *social justice* theme, which can overlap with socialist versions of just transition and climate justice campaigns (Scottish Trades Union Congress, 2010). Variations of these narratives and combinations of them are drawn on to support underlying discourses that serve particular class interests or shifting alliances of classes struggling for hegemony. For example, the combination of apocalyptic narrative in the mass media with individualized lifestyle-change narratives has provided the impetus for communitarian narratives (Hopkins, 2008; Segnit & Ereaut, 2007), as well as calls to direct action at the CCA (Plows, 2008, p. 96). We explore three discourses representing distinct class interests before examining the possibility of a discourse of transformation.

DISCOURSES AND CLIMATE CHANGE

Social-democratic discourse of climate change

Social democracy was the hegemonic political discourse, particularly in the UK, from the end of the Second World War until its influence started to decline in the late 1970s. Combining a welfare state with Keynesian state intervention in the economy, the significance of social democracy lay in its compromise between, on the one hand, the interests of local, national, and transnational capital accumulation and, on the other, the interests of the working class and the poor.

Although social democracy has been on the retreat, the discourse remains significant in Europe and reflects particular stages of hegemonic struggle between different classes. In Scotland, Jack McConnell, the Labour first minister of the Scottish Executive (government), gave a speech in 2002 that committed his government to environmental justice by incorporating the environment of the poor into a discourse of social democracy.

> The people who have the most urgent environmental concerns in Scotland are those who daily cope with the consequences of a poor quality of life, and live in a rotten environment—close to industrial pollution, plagued by vehicle emissions, streets filled by litter and walls covered in graffiti ... In the late 20th Century the big political challenge—and the greatest success I believe—for democrats on the left of centre was to develop combined objectives of economic prosperity and social justice. I believe the biggest challenge for the early 21st century is to combine economic progress with social and environmental justice.
> (Scottish Executive, 2002)

Moreover, this vision of environmental justice includes a global dimension, in which the unequal cause and unjust impact of climate change are extensions of the argument for social democracy. However, uneasy compromises with a resurgent neoliberalism have added to the decline of social democracy as a coherent ideological project (see Scott & Mooney, 2009). In Scotland this compromise was evident in environmental-justice policy, which influenced social policy but made no impact on economic policy and therefore did not directly address the causes of environmental injustice (Scandrett, 2007, 2010).

Neoliberal discourse of climate change
The aim of neoliberalism is to reestablish the power of the capitalist class, unhindered by compromise with any other class (Harvey, 2005). The

Poverty, Protest, and Popular Education

discourse combines a laissez-faire free market with a rhetorical commitment to reducing the role and size of the state, although in practice this applies only to certain sections of the state—those parts considered to be in competition with private capital.

There is no single neoliberal discourse on climate change. At one extreme, it simply denies that climate change is occurring, treating the scientific discourse as a constraint on business freedom and economic growth to the advantage of professional interests. In another version, the reality of climate change is accepted, yet it is regarded as irrelevant to the principal task of capital accumulation. For example, the chair of a major UK bank in the years prior to its spectacular 2007 collapse stated, "The richer they [the 'developing' countries of the global South] get, the less weather-dependent their economies will be and the more affordable they will find adaptation to climate change" (quoted in Gray, 2010, p. 51). More sophisticated neoliberal discourses focus on weak sustainability, assuming that financial capital can easily substitute for natural capital, and allocating property rights to elements of the carbon cycle (carbon dioxide emissions, carbon sequestration capacity), thereby creating a free market. The result leads to a trade in carbon derivatives, future carbon markets, and risk, with the capacity for subprime carbon credits ("futures contracts to deliver carbon that carry a relatively high risk of not being fulfilled"; Friends of the Earth U.S., 2009).

Though it is the dominant policy discourse throughout much of the world, neoliberalism has yet to develop a convincing narrative on climate change. This is perhaps part of the reason for the near universal failure of climate change policy, which remains a constraint to the accumulation of capital and to the power of the capitalist class. For this reason an alternative discourse of climate policy has received approval in some policy contexts in which the state retains an interventionist role in supporting the accumulation of capital—namely, ecological modernization.

Ecological modernization discourse of climate change

Ecological modernization (EM) attempts to integrate ecological interests into market forces by such means as allocating property rights to natural commons (e.g., water privatization, enclosure of the atmosphere in carbon trading) and socioenvironmental knowledge (e.g., biopiracy, dispossession of indigenous technology); constructing commodities from waste (e.g., energy), or through stimulating quasi-markets (e.g., pollution-permit trading) and through state-manipulated price feedback (e.g., fuel tax, cap and trade) or contingent valuation (e.g., willingness to pay).

Dryzek, Downs, Hernes, & Schlosberg (2003) differentiated between weak EM, which makes "capitalism less wasteful within the existing framework of production and consumption" and its strong variety, which would "democratise the state ... by including environmentalists in the core, creating the green state" (p. 167). EM tries to reconcile environmental conservation with economic efficiency by cooperation between governments, business, scientists, and reform-oriented environmentalists.

Whereas strong EM is critical of dominant policy paradigms, weak EM tends toward neoliberal market mechanisms (Dryzek, 2005). There is, in some respects, a triad of sustainability, social justice, and economic concern that makes for a complicated discursive mix. The influence of this on policy in Scotland takes place largely through its presentation of supposedly neutral technological narratives. For example, Scotland's Zero Waste Plan (2010)

> is probably the most consistent example of the use of EM thinking so far, employing state-manipulated market-led "closed loop" management of resources (that is, treating waste in one part of the production process as a resource for another part). The policy approach is based on maximising the economic value of waste materials in order to stimulate business activities in reuse, recycling, and recovery of energy.... This combination of technical and

> entrepreneurial activity is indicative of EM and serves the interests primarily of business and the professional class.
>
> (Scandrett, 2012, p. 253)

Just as the welfare state embraced a contradiction between class interests that eventually led to its fissure in the 1970s, so too is there a contradiction in EM:

> On the one hand, ecological modernisation provides a common discursive basis for a contested rapprochement between [the environmental movement] and dominant forms of political-economic power. But on the other, it presumes a certain kind of rationality that lessens the force of more purely moral arguments ... and exposes much of the environmental movement to the dangers of political cooptation.
>
> (Harvey, 1996, p. 378)

Harvey suggested that the environmental justice movement has influenced EM to the extent that "some sort of configuration has to be envisaged in which ecological modernization contributes both to growth and global distributive justice simultaneously" (p. 379). As shown in what follows, this has been the case, in some sectors, of ENGOs and environmental activists who have attempted serious dialogue with environmental justice struggles and other working-class movements in order to build a dissenting discourse (Scandrett, 2007).

Socialist transformation discourse of climate change

A dissenting discourse with the capacity to challenge capital accumulation must include the contribution from subaltern classes and, in particular, the victims of climate change and other environmental injustices (Bellamy Foster, 2010). Environmental justice movements come from a wide range of contexts throughout the world in which people who are directly affected by environmental damage mobilize in community or workplace campaigns and social movements of resistance. Those directly involved tend to be communities whose members are poor, working class, racialized minorities, indigenous peoples, politically disenfranchised, and geographically isolated.

Environmental-justice campaigns form around issues of social justice: racial segregation, land rights, workplace hazards, industrial compensation claims, poverty, and urban decay (Faber & McCarthy, 2003). The movements are often distinct from mainstream ENGOs whose membership largely comprises educated middle-class professionals, although there are attempts to build alliances between them. Martinez-Alier (2002) argued that environmental justice movements emerge when communities whose poverty and social position deny them leverage in the market begin to resist the cost-benefit analysis of development. Such movements develop incommensurable languages of valuation to the cost-benefit analysis of economic development, leading to discursive social and political conflicts.

Climate change is possibly the ultimate conflict in the conditions of production (O'Connor, 1998), in which capitalism destroys the environment that makes capital accumulation possible and leads to an irresolvable crisis. In this analysis, social movements are likely to emerge among the victims of climate change with common interests in challenging its causes.

> The main historic agent and initiator of a new epoch of ecological revolution is to be found in the third world masses most directly in line to be hit first by the impending disasters.... They too, as in the case of Marx's proletariat, have nothing to lose from the radical changes necessary to avert (or adapt to) disaster.
> (Bellamy Foster, 2010)

Harvey (2006) argued that climate change involves accumulation by dispossession in which the global commons and significant geoecological cycles that make life possible on earth are appropriated in the interests of capital accumulation. In which case, accumulation by dispossession entails a very different set of practices from accumulation through the expansion of wage labour in industry and agriculture. The latter, which dominated processes of capital accumulation in the 1950s and 1960s, gave rise to an oppositional culture (such as that embedded in trade unions and working class political parties) that produced the social democratic compromise. Dispossession, on the other hand, is fragmented and particular.... It is hard to oppose all of this specificity and particularity without appeal to universal principles.

(Harvey, 2006, pp. 52–53)

Fragmented forms of militant particularism can therefore unite only against the causes of climate change through access to abstract knowledge, which is usually developed by the professional class.

However, the professional class depends on high levels of public expenditure in education and public services, which is being undermined; this might result in its decomposing into the working class from which it has emerged over the past few generations. The ascendancy of the professional class contributed to the flowering of social movement activities in the postwar welfare-capitalist countries. They are the scientists, policy developers, and public sector workers who often constitute the membership of the ENGOs. Unlike most environmental-justice movements, the campaigns against climate damaging activities are led by the professionals in the ENGOs in the global North and their more recent offshoots in the direct action environmentalists (Seel, Paterson, & Doherty, 2000). Action to challenge both the causes of climate change and the powerful social interests that profit from current arrangements must make alliances between this class and the poor.

We propose that popular education can develop and reinforce discourses of dissent and alliance building. By *popular education* we mean the deliberate educational techniques devised originally by Paulo Freire and adapted and developed by radical educators and grassroots social movements (Crowther, Galloway, & Martin, 2005; Crowther,

Martin, & Shaw, 1999; Freire, 1972; Hope & Timmel, 1984; Kane, 2001). This educational approach refers to activities to create dialogical exchanges between forms of knowledge and culture—in particular, those informing climate change narratives and discourses and the material interests of the subaltern class or group. "Knowledge from below" is not necessarily better or correct, but it widens the opportunities for generating knowledge that is relevant for engaging people in social change. Such engagement includes meetings, networking, correspondence, electronic communication, publication exchange, joint campaigns, common struggles, and so on, all of which can include a component of dialogical collective learning.

Case studies

The following three cases involve genuine attempts at dialogue between environmental activists and working-class communities directly affected by climate change. Some of these use popular education; in others, elements of popular education methodology can be discerned. These cases are presented because they provide examples of attempts to build on climate change discourse with subaltern classes and expose contradictions and problems for practice, the analysis of which can lead to further work.

Friends of the Earth Scotland's environmental justice campaign

In 1999 Friends of the Earth Scotland (FoES) used the opportunity of political devolution in Scotland to launch a campaign for environmental justice that explicitly linked local environmental injustices with the globally unequal distribution of resource consumption, especially fossil fuels, and their resultant waste stream, climate change. The campaign employed a strapline—"no less than a decent environment for all, with no more than our fair share of the earth's resources"—that became widely influential among NGOs (see Agyeman, 2005; Boardman, Bullock, & McLaren, 1999; Dunion, 2003; Dunion & Scandrett, 2003; McLaren, 2003).

In this context, FoES's practice shifted to include active support for working-class or poor communities directly affected by environmental pollution and degradation, explicitly drawing on popular education (Agents for Environmental Justice & Scandrett, 2003; Scandrett, 2000; Wilkinson & Scandrett, 2003;). Over a period of six years, FoES engaged in sustained dialogue through structured educational programs with communities fighting against waste landfills, opencast coal mines, quarries, polluting industries, fish farms, poor-quality housing, and workplace hazards. Popular education was also used to generate dialogue between campaigners in Scotland and activists who are victims of the fossil fuel industry elsewhere in the world. Several exchange visits took place involving Ecuadorian activists, including some from the industrial city Esmeraldas, where a leak from an oil refinery led to a fatal explosion, and representatives of the Cofan nation, an indigenous people whose ancestral land was expropriated, exploited, and polluted by oil extraction (Scandrett, O'Leary, & Martinez, 2005).

As a direct result of this campaigning work, the Scottish Executive (now called the Scottish Government) included environmental justice in its policy discourse by positioning climate change within narratives of inequality and social justice as a logical extension of social democracy, subject to neoliberal economic growth. Even as environmental justice was incorporated into policy thinking, however, its meaning was diluted toward the management of what Curtice, Allaway, Robertson, Morris, Allardice, and Robertson (2005) have called "environmental incivilities": the pollution of social space through antisocial activity, such as littering, fouling streets, and graffiti (Scandrett, 2007). By contrast, and through ongoing dialogue with directly affected communities, FoES's narratives increasingly focused on the common causes of diverse environmental injustices reflected in wider social and economic injustices.

However, FoES's commitment to environmental justice reflected the tension between developing community-based campaigns and serving its own membership. Like those of many environmental organizations, its members are drawn predominantly from a middle-class, highly educated section of the population that might not readily identify with environmental-justice issues affecting poor, working-class communities. An online questionnaire to FoES members completed by 174 respondents in 2007 revealed a social profile in which over 77% worked in white-collar occupations, such as professional and administrative work, and over 65% were educated to degree or postgraduate degree level. Research among those who participate in online campaigns illustrates some of the complexities of discourse alliance in ENGOs (Scandrett, Crowther, & Hemmi, 2009). Our survey showed that 78% of the sample prioritized campaigns on "global climate change," whereas 47% prioritized "poverty and pollution," a significant figure almost certainly the result of the environmental-justice campaigns conducted over the previous decade. However, specific environmental-justice issues that predominantly affect people living with poverty were given low prioritization: local planning issues (7%), fuel poverty (5%), incinerator campaigns (0%). Moreover, environmental issues with negligible associations with environmental justice received higher prioritization: renewable energy technologies (48%), wildlife conservation (38%), population growth (24%). FoES members could support environmental justice in the abstract but without a major shift in the focus of their practice. Though the environmental justice campaigns involved a significant attempt to build alliances between these classes and to change the ENGO discourse toward the interests of the working class (Scandrett, 2007), its success was limited.

In 2007 the Scottish Labour Party lost the election to a minority Scottish National Party government that abandoned discourse on environmental justice, and by this time FoES had begun to direct its campaigns more explicitly to members' interests of lifestyle environmentalism, green technology, and policy-focused lobbying. Climate justice was diluted to climate change. It should, however, be noted that subsequent continued ENGO lobbying on the issue of climate justice in Scotland has recently resulted in the creation (May 2012) of a Scottish Government Climate Justice Fund for climate adaptation in developing countries. At the time of writing, the fund is meager (£1 million), but it represents a small discursive victory on the part of ENGO lobbying. It also adds to a policy commitment that has been established in Scotland through resourcing community action addressing climate change issues through the Climate Challenge Fund (CCF).

Poverty and the Climate Challenge Fund

Communitarian narratives play an important role in the Scottish policy context. The CCF has, at the time of writing, provided £37.7 million to over 500 community-level carbon-reduction projects across Scotland, and a further £30.9 million will be made available between 2012 and 2015 (Scottish Government, 2012). The fund was the initiative of Green Party members of the Scottish Parliament in return for their support for the minority Scottish Nationalist Party government in 2007. Community projects were able to take advantage of the fund and deliver projects in ecological renovation, local food production, heating, transport, and waste—projects that both benefited local communities and delivered reductions in carbon dioxide emissions. Some of the beneficiaries of the fund also integrated climate change into community development work designed to tackle poverty.

Interviews were conducted with a community education worker in one such project in a working-class community with a long history of poverty and deprivation. For this individual, the purpose of engaging with the CCF was partially to respond to the changing policy agenda in order to access funding for antipoverty work and partially to build on community education practice in order to generate insights on climate change from the perspective of those struggling against poverty:

> We've tried to tackle issues to do with poverty in this community for a long time ...[W]e were very straight with the Climate

> Challenge Fund right from the start that we would be taking a poverty [focused] approach to this ... So in a sense, we're doing a classic community education thing which is trying to use the policy agenda to bring about outcomes that the community would value...
>
> [People from the community] may come in because [they] want to save a bit of money. But you try to actually widen their understanding of climate change through that: where that actually leads to, why cut down CO_2 emissions, what impact that will have, and the whole broader political debate. Because otherwise, you're actually not crediting people with intelligence.... It's about ownership of knowledge.... You know, how do you engage critically with the world.
>
> (Anonymous community worker, personal interview with McGregor, May 25, 2010)

This is an explicitly dialogical educational agenda involving responses to policy opportunities presented by climate change initiatives, with activities that generate both material benefits and educational opportunities for a working-class community struggling against poverty. It also exposes opportunities that community workers are able to exploit and that reveal the role of powerful discourses in everyday practices. In constructing questions for a community "carbon baseline" survey, the community educator recalled an incident that illustrates, in a micro context, the discursive nature of power and the practices which constitute it:

> We did a [carbon] footprint survey with another organization that's worked with climate change issues for a long time ... They designed the survey.... We then looked at the survey and ... insisted we include questions like "How many hours a day is your heating on for?" because a lot of people in this area are unemployed ... Also, I wanted to ask, "In the winter do you feel your home is warm enough?" ... [With] those two questions, you immediately get into a fuel poverty issue.

Poverty, Protest, and Popular Education

> [W]hen the report came out we weren't happy with it: They'd said ... they're heating their homes more than the average community. And I would say see you've reached the wrong conclusion ... Their energy bills are high because of the predominance of electric heating and prepayment meters ... At one point the organization said to us, "Look, we're experts," and I said, "This is not just for you; we're seeing this as a bit of knowledge for the community to use for their own ends" ... We need to be able to use this as ammunition to fight, for people to be able to fight to get the changes—this is a great opportunity for people to make arguments about the state of their homes where they've never been able to win any ground in the past.
> (Anonymous community worker, personal interview with McGregor, May 25, 2010)

The problems of long-standing poverty and disempowerment are compounded by the vested interests of powerful groups engaged in generating climate change discourses.

These contradictions illustrate that mutual learning in encounters between different interests—a local community development trust and a professional sustainability consultancy—can really be understood as an episode of local "policy epistemics" (Fischer, 2000). That is, meaning is constructed in a dialectical relationship between local people and experts in which the educator's aim is to make the assumptions of the dominant discourse visible and negotiable. Furthermore, it is important that "sites of discursive resistance" (Flowerdew, 2008, p. 205), such as this one, are documented so that change agents attempting to foster dialogue between climate change and antipoverty initiatives can develop a better understanding of where opportunities lie and what the obstacles are. Reinforcing this particular account, a commissioned review of the fund (Brook Lyndhurst & Ecometrica, 2011, p. 114), suggests a reorientation of the strategic aims of the CCF beyond funding projects that are able to demonstrate short-term minor carbon impacts, toward supporting projects aimed at "encouraging sustainable living and capacity building for action."

Despite recognizing and responding to the opportunities for constructing alternative discourses based on the experience and knowledge of working-class groups struggling against poverty, there is a real sense of the practical difficulties of engaging with this issue. With a long history of disempowerment, the potential for engaged political struggle for social transformation is limited.

> I think we're at the real initial stages of trying to engage people. It's potentially part of a social movement, but it's probably not there yet. There's potential to build that. To me what's worrying is the fact that a message like climate change for a community like this is going to take a long time to work on ... it's not really got enough of an immediacy for people and even the issues that do have an immediacy for people don't necessarily get people going. So how can you get something like climate change really getting people going?
> (Anonymous community worker, personal interview with McGregor, May 25, 2010)

Recognizing and responding to the opportunities for constructing alternative discourses is an arduous path strewn with real practical difficulties. However, some sections of the environmental movement have moved away from trying to exploit the ambiguities of policy initiatives, or from becoming trapped in their potential limitations, and have taken up direct action.

Direct action environmentalism: Camps for climate action

DA environmentalism grew during the 1980s and 1990s as a response to the rising professionalization of the mainstream ENGOs and their apparent lack of success against neoliberal governments in the United States and the UK (Seel, Paterson, & Doherty, 2000). Drawing inspiration from anarchism, deep ecology, animal rights, and the peace camps,

DA environmentalists developed new forms of autonomous organization and confrontational political protest that focused on individual responsibility to directly prevent damage to the environment. The demography of the DA environmentalists is disproportionately young and, like mainstream environmentalism, well educated and professional (Rootes, 2004, p. 619). An important focus of their environmentalism in the UK since 2007 is an annual CCA, a few days during which hundreds of activists occupy land close to a source of pollution (in 2010 this took place at the headquarters of the Royal Bank of Scotland [RBS], the partially publicly owned bank notorious for financing ecologically and socially destructive developments).

RBS finances the extraction of fossil fuel from tar sands in Canada, a process that not only contributes significantly to climate damage but also dispossesses people of the First Nations, on whose land the tar sands are developed. The CCA claimed to be acting in solidarity with these First Nations, representatives of which were present. When windows were smashed in an unsuccessful attempt to gain access to RBS offices, the First Nations representatives expressed concern about the damage to property. This led to a wide-ranging debate at the camp; views ranged from giving an unconditional apology to defending tactics that had been developed as part of the DA repertoire, and to a desire for more effective communication between the two cultures of protest.

Another comparable conflict of cultures of protest occurred when a group of DA environmentalists associated with Earth First! and CCA became aware of the growing expansion of opencast coal extraction in Scotland. For them, there was a clear connection between climate change and the extractive fossil fuel industry directly affecting local working class communities:

> Despite the fact that the "climate change movement" or "radical environmental movement" was growing, it was ... turning further away from achieving the kind of social change that will be necessary to bring about any kind of revolutionary ecological society.

> "The movement" wasn't really talking about what really affects [working-class] people, and increasingly our politics dealt with abstract carbon-counting, made worse by our obsession with the financial crisis.
>
> (Coal Action Scotland, 2009, n.p.)

This led to a series of "digger diving" activities (immobilizing digging machinery) at various opencast mine sites and eventually to the establishment of a Solidarity Camp at Mainshill, the proposed opencast site in South Lanarkshire. In a report of the action incorporating advice to activists, Coal Action Scotland (ca. 2009) commented:

> Many community campaigns focus only on the legal process, or pin their hopes on stopping a project at the planning stage. They know better than anyone the endemic corruption at local council level but many community groups will want to have exhausted all the legal options before supporting an occupation—although you may disagree with leaving direct action until all else has failed, be sensitive to this. As trust is built up with the communities, hopefully direct action won't just be seen as a last resort to be taken by intrepid eco-activists, but as an integral part of a campaign and an expression of the communities (sic) right to self-determination.
>
> (Coal Action Scotland, 2009, n.p.)

The solidarity campers put time and effort into building relationships with the local working-class community through informal contact in the pub and other community spaces. The community members, it seems, were generally glad to have the support of committed activists, invited them in for tea, dropped by the camp and contributed food and materials for their temporary shelters. Much of this mobilizing support is very informal and unstructured and lacks a concerted attempt to achieve common platforms or collective dialogue. Because of the illegal nature of much of the activity, the DA environmentalists' mode of operation involves individuals and small groups carrying out actions secretly, within the parameters of shared codes of conduct but without even other activists' knowing. It was in the context of a piece of secretly conducted "pixie action" (sabotage) that tensions between the activists and the community erupted.

> In the summer, during The Camp for Climate Action at Mainshill some people snuck out in the night and dismantled the conveyor belt at Glentaggart opencast mine. This was the first pixie action to be reported since the camp had arrived and it provoked an interesting and difficult discussion between campers and the local anti-opencast campaigners. At the camp we were excited by the news of the action and generally pleased that it had happened. But, in a large meeting at the camp some of the locals told us that they were unhappy with this kind of action—that it increased the amount of lorries transporting coal by road in the area, that we'd crossed a line and that if it happened again we wouldn't be seeing them at the camp any more.
> We'd pissed off the very people that we were there to support and it didn't feel good. This was a hard blow to our enthusiasm to push our limits and step-up our tactics to take the fight direct to those waging destruction. However, our relationship with the locals moved on from this point with continuous communication and although it is difficult to have frank discussion about anonymous actions I do believe that the supportive locals came with us on a journey of radicalisation that made our resistance stronger.
> (Coal Action Scotland, 2009, n.p.)

This conflict arose between two cultures of militant particularism. The traditions, codes of conduct, and nomenclature of the DA environmentalists have grown through praxis and debate during actions and camps, on campuses and in squats, online and in the samizdat publications of the movement. This is a culture outside that of conventional working-class community action, which has its own traditions of collective accountability, a critical respect for legal and state processes, and an intimate knowledge of the local area and workings of opencast mines. The locals also had more to lose in the long term and bore the consequences of increased lorry movements and association with criminality.

Some DA environmentalists have gone a little further in understanding how dialogue between cultures develops. In 2009 DA environmentalists associated with the anti–airport expansion group Plane Stupid organized a "convergence" with campaigners against fuel poverty in Clydebank, a working-class community under the flight path of the Glasgow airport, for which expansions were being considered at the time. These fuel poverty activists were victims of the oil industry in several ways: Living in poor quality housing, they spent a disproportionate amount of their low incomes on fossil fuels to heat their homes and water in order to provide a basic level of comfort and hygiene taken for granted in most European houses. At the same time, they suffered the constant noise and gas emissions of low-flying airplanes taking off and landing close to their community.

The convergence used popular education to generate discussions on tackling fuel poverty without contributing to climate change: The Clydebank activists' main focus was supplementing their social security benefit income to pay for increased heating, whereas for the climate activists, structural improvements to housing were necessary. The role of the state was also a source of debate: For the fuel-poverty campaigners, the state was the source of their income and the focus of their campaigning, and they asserted their democratic right to both; for the DA environmentalists the state was the enemy and inevitably corrupt.

At the time of writing several initiatives in process use popular education to construct dialogues between DA climate activists and working-class interests—for example, a preliminary meeting between a small group of DA activists and trade union representatives, including the shop stewards' convener at the nearby Grangemouth oil refinery, who represents over 50,000 oil industry workers. This seeks to build on the UK trade union initiative for a million climate jobs, based on social-democratic/socialist narratives of major state investment for a national climate service (Campaign Against Climate Change, 2009). There are also plans for a "toxic tour," organized by DA activists and other environmentalists to link communities in central Scotland with the negative experience of climate affecting industries.

Conclusion

> It is no use simply saying to South Wales miners that all around them is an ecological disaster. They already know. They live in it. They have lived in it for generations. They carry it with them in their lungs.... But you cannot just say to people who have committed their lives and their communities to certain kinds of production that this has all got to be changed. You can't just say: "come out of the harmful industries, come out of the dangerous industries, let us do something better." ... If [environmentalists and the working class do] not really listen to what the other is saying, there will be a sterile conflict which will postpone any real solutions, at a time when it is already a matter for argument whether there is still time for the solutions.
> (Williams, 1989, p. 220)

As Williams pointed out, dialogue between environmental and class-based struggles is essential to the search for just solutions to ecological problems. There remains the possibility of constructing new hegemonic blocs of class interests, incorporating fragmented groups that share the experience of being exploited by the fossil fuel industry, in dialogue with the science-based discourses of climate change used by scientists, environmental campaigners, and policy makers of the professional class. Through such processes, the connection between fragmented forms of oppression and climate change can be made, and there is a capacity for generating an alternative, transformative, climate change discourse. This project requires understanding the relationship between discourses and the alliances of class interests that make up hegemonic blocs at any particular stage in the process of social change.

It is in the interfaces between the professional class and the subaltern classes that popular education can make a contribution. Examples of engagement between environmental discourses of climate change and working-class communities are small but significant. Case studies from ENGOs, community development, and DA environmentalism illustrate the potential and the difficulties of constructing dialogue, and they also provide evidence that popular education provides useful tools for this process.

Current attempts at popular education are small and fragile and swimming against a powerful tide, but history has shown that the momentum for change can build suddenly and quickly. We can only advocate continuing work in this area of popular education allied with struggle, in the hope that transformation occurs before human societies and the ecosystems of which they are part are damaged irreversibly.

Endnotes

1. We use this term with caution: Lahsen's (2007) neo-Gramscian research on power relations between climate scientists in the majority world and the global North reveals the notion of a benign and collegial transnational intersubjectivity in the creation of IPCC discourse to be mythical. Citing Hass, Lahsen argued that scientists from "less developed countries" made up only "17.5% of the scientists producing and reviewing the IPCC's Third Assessment Report" (Lahsen, 2007, p. 180). Lack of trust between scientists in the majority world and global North was argued to be contingent on "national level interpretive biases" to do with "democratic norms, ethics and equity" (Lahsen, 2007, p. 189).
2. A caveat must be acknowledged—namely, the principal discourses on climate change that are addressed concern the release of carbon dioxide through the burning of fossil fuels. This is only one component of the complexity of climate change, which includes carbon loss from modern, intensive, and chemically saturated agriculture (between 1980 and 2003, an estimated 12%–15% of total carbon was lost from soil through arable agriculture in England and Wales, according to Bellamy, Loveland, Bradley, Lark, & Kirk, 2005) and the emissions of other greenhouse gases, such as methane from organic waste and cattle farming.

CHAPTER 11

COMMUNICATING FOR SUSTAINABLE CLIMATE POLICY

*Tarla Rai Peterson
and Anabela Carvalho*

The preceding chapters critically examine climate change communication in a variety of venues, searching for signs of life in the politics of climate change. They identify and explore varying forms of critique and contestation that challenge the status quo. Each one demonstrates in multiple ways that communication not only provides a channel for exchanging information and meaning but also constitutes possibilities for climate change politics. Here, we briefly summarize the main points made in chapters 2 through 10 and then suggest potential opportunities for connections between citizen communication and climate change politics.

Citizen communication about climate change often remains within the limits of existing institutional structures and focuses on individual lifestyle choices. Chapters 2, 3, and 4 focus on individual behaviors and examine the potential for persuading people to change those behaviors. In chapter 2, Ryghaug and Næss examined multiple ways that climate change and people's perceived powerlessness to solve it are discursively organized and (re)negotiated in the domestic context of daily life. They showed that understandings of climate change and of climate change politics develop out of relations between media discourses, political discourses, and mundane living conditions. They noted that apparent contradictions between the alleged urgency of the problem and obvious political inaction regarding climate change encourage political disengagement among Norwegian citizens.

Schweizer and Thompson examined possibilities for encouraging behavior change among U.S. citizens, focusing on the U.S. National Park Service's largely unrealized opportunities to communicate information about climate change to a relatively friendly audience. They examined the climate change messages developed and disseminated by the U.S. National Park Service and suggested that a persuasive campaign that meshes with cultural values and beliefs of park visitors could encourage more climate-friendly behavior, including public pressure on political leaders. Despite Schweizer and Thompson's optimistic outlook, however, climate change is so spatially and temporally vast that the limitations of individual behavior change seem daunting. Further, both Schweizer and Thompson's and Ryghaug and Næss's analyses seem predicated on an assumption that climate policy will follow changes in public opinion and that individual behavior change can make important contributions to climate change mitigation. This seems somewhat naïve given that, as noted in chapter 1, polls have long indicated that a majority of the public (across multiple nations) thinks climate change is a serious problem. Despite these results, few nations have developed and implemented strong climate policies.

Whereas Schweizer and Thompson discussed the potential for materializing climate change through guided experience with natural features, Hughes explored another technique for communicating climate change. She examined how filmmakers have materialized this otherwise abstract concept through soaring landscape images. Hughes suggested that by reconstituting climate change through the eye of high-angle extreme long shots, filmmakers provide a space in which individuals can experience the immediacy of climate change without reducing it to something that can be addressed by switching to fluorescent lightbulbs. Although Hughes's references to strategic communication also imply a social-marketing approach to public engagement, her analysis focuses on constitutive, rather than instrumental, dimensions of communication. Along with Ryghaug and Næss, she noted that media discourses have the potential to reconstitute climate change for large numbers of the public. How those new meanings translate into policy or significant civic mobilization, however, remains uncertain. Further, she noted that even as they target different audiences, the films she analyzed locate the potential for change outside the traditionally recognized corridors of political power. In fact, given its pairing of precisely honed data and emotionally sublime images, the cinematic use of extreme long shots may offer an iconic example of agonistic politics.

Chapters 5, 6, and 7 shift attention to public participation in climate change politics. They demonstrate concern with how established institutional mechanisms enable (or disable) various publics to voice opinion, perspective, and interest in hopes of improving climate change policy. They also look at different pathways to communicate and enact public participation. Carvalho and Gupta's analysis of communication reports submitted by signatories of the United Nations Framework Convention on Climate Change (UNFCCC) shifts attention to the more formally recognized political realm. For this analysis they adopted Dietz and Stern's (2008) traditionalist definition of public participation as "organized processes adopted by elected officials, government agencies, or other public- or private-sector organizations to engage the public" (p.

1). They showed that states refute their commitments toward promoting public participation in climate politics and construct citizens as recipients of information rather than political subjects. The extent to which public participation in policy processes may help bring about appropriate climate change policies is unknown. Nevertheless, it is plausible that civic input into governmental decision making and dialogical mechanisms between state and society would produce acceptance for the policy measures required to address climate change. Carvalho and Gupta's analysis of six country cases shows that by excluding references to the topic in their official reports, governments turn public participation into a political nonissue.

In chapter 6, Jönsson demonstrated that social media have enabled citizens to develop alternative forms of governance. Jönsson has joined other media scholars who credit the Internet with a revival of democracy by highlighting its role in social mobilization (e.g., Atton, 2004; Dahlberg & Siapera, 2007). Yet we remain skeptical of claims that the Internet alone will bring society closer to strong democracy. Though it recognizes digital divides, such as those along age, gender, and income lines, Jönsson's chapter makes a persuasive case for Facebook and Second Life as epicenters of a new, transnational deliberative politics that has the potential to influence national and international policy. Rather than assuming that social media automatically enable more and better political participation, this chapter argues that social media provide different venues for dissent, including some that are less risky than embodied protest. For example, participation in Facebook and Second Life allow citizens to put forth ideas in an editorial-like form without the spatial and temporal constraints of embodied protest—and allows them to do so in a context where they can control how much personal, identifying information to divulge.

In chapter 7, Feldpausch-Parker, Parker, and Peterson examined 350.org, a climate change nongovernmental organization (NGO) that had hoped to influence international climate policy via the Copenhagen

Accords. The organization's stated goal of influencing international climate policy fits neatly within the framework of deliberative democracy. Both central campaign coordinators and local event organizers had hoped to influence the policy-making process. However, the notion that a motley band of students, local community activists, and other "ordinary" citizens could organize a global set of events is decidedly nontraditional. They were not averse to using conventional political channels when available, and like participants in Second Life, they used the Internet to orchestrate resistance to official inaction on climate change. By weaving Internet and embodied action together, they attempted to mobilize political resources and heighten official recognition that climate change demands radically new policies. Their nonconformist positioning gives the initiative an element of agonistic politics.

In chapter 8, Polli shifted the focus from direct participation in existing bureaucratic mechanisms for developing climate policy by exploring ways that activist art can promote agonistic pluralism. She argued that technomanagerial constructs, such as emissions markets and carbon trading, further ingrain the traditional responses that have led to the current debacle. She lamented that in a world where everything has been privatized, politics has become an empty signifier. Offering art as a powerful curative to the political malaise associated with modification of the atmosphere, Polli described ways it can destabilize hegemonic understandings of earth systems, including those associated with climate change. Polli's artists mobilized political resources to foment the dissensus needed to give voice to those who have been silenced by current political hierarchies, and her analysis moves from a focus on public participation within established venues to an exploration of grassroots political mobilization.

In chapter 9, Gunster examined ways that alternative media contribute to an agonistic politics by framing the politics of climate change differently from mainstream media. As Gunster noted, the media produce and disseminate multiple, sometimes conflicting, discourses on climate change. By providing spaces where readers and viewers can confront disparities of knowledge, values, and power that mainstream media often gloss over, alternative media facilitate public confrontation with currently hegemonic projects and highlight their influence over climate change policy. Scandrett, Crowther, and McGregor followed Gunster's media analysis by characterizing current climate change communication as an exclusionary process that silences the voices of large segments of society. They demonstrated that the political dimensions of climate policy clearly emerge as issues of class and social justice. They argued that a radically democratic climate politics could be constituted via particular forms of communication, education, and social relations. Their conclusions suggest that communication can provide the beginnings of a response to Bullard, Johnson, and Torres's claim that

> as we search for ways to rectify global climate change, we desperately need the input of the populations most likely to be negatively affected: people of color and other poor people in the North and in the developing countries of the South.
> (Bullard, Johnson, & Torres, 2005, p. 292)

Scandrett, Crowther, and McGregor noted that constituting a climate politics that extends beyond the elite segments of society that have thus far controlled the issue requires a radical reorientation that must include the possibility of direct confrontation needed for development of an agonistic politics on climate change. According to the International Climate Justice Network (2002), "the biggest injustice of climate change is that the hardest hit are the least responsible for contributing to the problem" (n.p.). The Intergovernmental Panel on Climate Change (IPCC, 2007c) indicated further that there is a growing climate divide between rich and poor communities regarding both where the effects of climate change are and will be the most acute and who is best prepared to address these effects, partially because "poor communities ... tend to have more limited adaptive capacities, and are more dependent on climate-sensitive resources such as local water and food supplies" (2007c, p. 359). Put

Communicating for Sustainable Climate Policy 313

simply, those in underresourced communities have neither the luxury of importing costly supplies nor that of leaving an area that is experiencing the effects of climate change. According to the International Climate Justice Network's (2002) Bali Principles of Climate Justice, "unsustainable production and consumption practices are at the root of [climate change] and other global environmental problem[s]" and "combating climate change must entail profound shifts from unsustainable production, consumption and lifestyles, with industrialized countries taking the lead" (n.p.). Moreover, these principles also define climate justice such that all people have access to necessities, such as sustainable energy. This means radical transformation. Only radical structural changes will enable such access, and those changes will not come without political clash. Through multiple case studies, this book highlights the constitutive relations between communication and different modes of political engagement with climate change. Though social marketing and public participation in institutionalized politics will continue to play leading roles in the politics of climate change, no foundation exists for assuming that climate policy will follow changes in public opinion; polls cited by several authors in this book have long indicated that a majority of the public worldwide is concerned with climate change, but few countries have put significant climate policies in place. The fundamental transformations that are needed can come about only through the third mode of engagement introduced in the first chapter: agonistic pluralism. Countering the depoliticization of prevailing technomanagerial solutions and opening up new policy possibilities for climate change requires a radicalization of democratic politics.

In the first chapter, we defined *the political* as engagement with processes of debate and decision making on collective issues in which different values, preferences, and ideals are played out and opposed. At a basic level, one must consider that people's political subjectivities determine their orientations to democratic participation. Because mediated political communication predominantly constitutes citizens into passive consumers (e.g., Lewis, Wahl-Jorgensen, & Inthorn, 2004) a transformative process of emancipation is needed to enable the political dimensions of public life to emerge. This transformation demands the creation of spaces for marginalized voices in both mainstream and alternative media, as well as in other communication spaces, where dissent rather than consensus is normalized. Such recognition could contribute to further civic mobilization, to engagement with social movements, and to incorporation of diverse views in political parties, among other political developments.

Political identities are not fixed but are relational and fluid. Rather than looking at citizenship as a legal status, Asen (2004) viewed it as a process and hence open ended and variable in its expressions. Noting that what matters is how people *enact* citizenship, he wrote of "fluid, multimodal, and quotidian enactments of citizenship in a multiple public sphere" (p. 191). Mouffe's rejection of essentialism in relation to political identities is consistent with Asen's process-oriented perspective and maintains that political identities emerge from the "contingent and pragmatic form of their articulation" (1993/2005, p. 7). The analyses in this book offer diverse contributions to rethinking citizenship regarding climate politics and the role of communication practices in promoting or hindering different forms of political engagement. Multiple tensions associated with perceived lack of citizen agency emerge in chapter 2 and elsewhere. Remarkably, however, despite existing barriers to politically relevant participation, several of the analyses indicate that citizens (individually and in organized groups) have been finding ways to relate to climate change that circumvent these limitations or enhance their agency and political efficacy—and have been finding that communication matters in those processes.

New information and communication technologies, and especially the Internet, offer an important potential to remake the political. In fact, many have hailed their role in the mobilization of social movements and the development of political activism (e.g., Atton, 2004; Dahlberg & Siapera, 2007). But appraisal of Internet politics is extremely variable; excited optimism coexists with profound disillusionment. Whereas some praise its transformation of the modes of social organization and political decision making, others point to the reproduction of capitalist logics and structures and to the dominance of meaningless content on the Web. Despite taking up a position critical of the "corporate and mainstream forms and uses of technology," Kahn and Kellner argued that information and communication technologies "have facilitated oppositional cultural and political movements and provided possibilities for the sort of progressive socio-political change and struggle that is an important dimension of contemporary cultural politics" (Kahn & Kellner, 2005, p. 76). These authors found political action opportunities in the integration of new modes of political activism, alternative forms of agency, and oppositional struggle via the Internet.

For example, 350.org began with the lofty goal of influencing international climate policy through the 15th Conference of the Parties (COP 15) to the UNFCCC in Copenhagen (December 2009). The group was part of an extensive social movement that developed around the goal of achieving an ambitious international agreement to curb greenhouse gas emissions. As Feldpausch-Parker noted in chapter 7, COP 15 was a severe disappointment to those seeking strong international climate policy. Yet 350.org has not lost its momentum. Since the October 2009 rallies leading up to COP 15, 350.org has coordinated 14 additional campaigns, and another seven campaigns are in the planning stages. Furthermore, the group has expanded to include other projects targeting public policies that directly impact climate change, such as energy development. Recent campaigns include "Put Solar on It," "Tar Sands Action – Stop the Keystone Pipeline," "Don't Frack the Delaware!," and "The 99% Spring"— in addition to continued climate action days. The organization claimed to have had a direct influence on the Obama administration's decision to deny a permit for the Keystone XL pipeline, which was slated to stretch from Alberta, Canada, to refineries in Texas. On November 6, 2011, the Keystone pipeline project mobilized participants for an image event that included encircling "the whole White House in an act of solemn protest" (350.org, 2012, n.p.). With attention drawn to the project, pipeline opponents were able to counter proponents' claims that the pipeline would further energy security by making more North American oil available to U.S. consumers. As public awareness grew that the tar sands were slated to be piped to Texas for refining because the finished product was intended for foreign export, political leaders began to question what had seemed to be a done deal. Bill McKibben, 350.org's default spokesperson, responded to Obama's decision:

> We've won no permanent victory (environmentalists never do) but we have shown that spirited people can bring science back to the fore. Blocking one pipeline was never going to stop global warming—but it is a real start, one of the first times in the two-decade fight over climate change when the fossil fuel lobby has actually lost.
>
> (350.org, 2012, n.p.)

Researchers who study phenomena that have demonstrable policy implications do not have the luxury of pursuing their research in neutral isolation. Industry, environmental NGOs, consumer advocates, and others will dissect research publications on climate science in the hope of finding a sentence or two that can be used to justify predetermined policy proposals. Those proposals may or may not be consistent with the findings or conclusions reached in any given paper. Perhaps this is one reason why some climate scientists have been willing to reach beyond their disciplinary expertise into the policy arena. Although James Hansen's claims regarding the biophysical components of climate change are consistent with current scientific consensus, his activism has caused discomfort among other climate scientists, who worry that his activities could sully the integrity of science. Even when referring to one of Hansen's recent publications in the prestigious Proceedings of the National Academy of Sciences, a colleague responded that "this isn't a serious science paper.... It's mainly about perception as indicated by the paper's title. Perception is not a science" (Gillis, 2012, n.p.). Although we agree that perception is not a science, neither is climate change.

The systematic study of either phenomenon, however, *is* a science. And perceptions of climate change constitute whatever political action is eventually enacted. If one assumes that continued human life on the earth is a good thing and that justice for all human beings is worth seeking, then the way one moves between climate science and climate policy matters. Climate change communication mediates that move, for better or worse, and understanding how it operates—whether through social media, public demonstrations, or formal policy reports—improves one's ability to craft a more sustainable and democratic existence.

The Internet alone is unlikely to bring radical and plural democracy, however. Though it is potentially more open than other communication media, the Internet also creates new divides along lines of age, income, nationality, and other factors. Moreover, it has been argued that the lack of personal contact and anonymity that characterizes it may result in less trust and responsiveness to social and political causes. Although there are advantages to the Internet as a space of political engagement, multiple such spaces should ideally coexist and interact: various types of online platforms and other new media, traditional media, alternative media (on and offline), the street, community centers, art galleries, workplaces, and other face-to-face and mediated spaces. Rather than discussing *the* public sphere, it may therefore make more sense to speak of a *plurality of public spheres* (cf. Breese, 2011). The level at which this may happen certainly matters. There are growing communicative practices that some may inscribe in the challenging notion of a transnational public sphere, but global climate change can and should also be brought down to other scales of (mediated) politics.

Mouffe argued that "a modern democratic theory must make room for competing conceptions of our identities as citizens" (1993/2005, p. 7). Societies are riddled with tensions and conflicting struggles, and citizens are infinitely diverse in their standings. Though not simple, the answers have to involve increased pluralism in addressing climate change at all stages of political life. But rather than expecting consensual deliberation, one should expect agonistic pluralism. As Mouffe has maintained, there will always be antagonism in social and political life; antagonism, she claimed, is in fact a condition of democracy, not an obstacle to it. For climate change as for any other issue, there is never going to be a unanimous solution or a final suture to the heterogeneous worldviews of political subjects. What are the means for and the implications of applying agonistic pluralism to climate change? Designing spaces and mechanisms of expression for a wide range of views on the problem, including those in disagreement with hegemonic discourses, would be required. Out of respect for liberty and equality, such arrangements would need to take account of the "different social relations and subject positions in which they are relevant" (Mouffe, 1993/2005, p. 71). These arrangements must enable confrontation of conflicting perspectives about how much one should risk and most likely allow to be destroyed in a world with increasing greenhouse gas emissions, about the worth of economic growth, about consumption, and about other social and political issues. Accommodating such a diversity of views on climate change maintains room for denialist positions but also opens space for those who view official policy as insufficient and inadequate, who strive for more substantial and ambitious goals. Obviously, this would pose new challenges. Still, defining (institutional) mechanisms to accommodate agonistic pluralism is a condition to produce better—if only temporary—responses to the enormous challenges posed by climate change.

The paths leading to such transformation are, to a large extent, still to be drawn. Yet we find hope in the numerous civic initiatives on climate change that have emerged in the last few years, several of which are examined in this book. Citizens, social movements, and nongovernmental organizations have been experimenting with alternative forms of communication, redefining the meaning of political action and participation. Asen's discursive notion of citizenship draws attention "to action that is purposeful, potentially uncontrollable and unruly, multiple, and supportive of radical but achievable democratic practices" (2004, p. 192). As shown by the analyses in this book, communicative practices, resources, spaces, and structures offer key opportunities for inventing alternatives that enable citizens to breathe new life into climate change politics and sanction policies that engage the multiple implications of climate change for contemporary society.

References

100 Million Voices for a Real Climate Deal in Mexico 2010 (Facebook group). (2010). Retrieved September 15, 2010, from http://www.facebook.com/#!/groups/212448202491/

350.org. (2009a). 10 min de luz de Velas [Caracas event post]. Retrieved October 24, 2009, from http://www.350.org/en/node/8341

350.org. (2009b). 350 from the Space Needle at 3:50 P.M. [Seattle event post]. Retrieved October 24, 2009, from http://www.350.org/en/node/7197

350.org. (2009c). 350 Madagascar [Fort Dauphin event post]. Retrieved October 24, 2009, from http://www.350.org/es/node/5625

350.org. (2009d). 350 prayer flags [Denver event post]. Retrieved October 24, 2009, from http://www.350.org/en/node/7380

350.org. (2009e). 350 Venezuela: Committed to a better climate [Cátia la Mar event post]. Retrieved October 24, 2009, from http://www.350.org/en/node/6196

350.org. (2009f). 350 Yemen March [Yemen event post]. Retrieved October 24, 2009, from http://www.350.org/de/node/5083

350.org. (2009g). Ann Arbor Fall ReSkilling Festival [Ann Arbor event post]. Retrieved October 24, 2009, from http://www.350.org/en/node/5860

350.org. (2009h). Community resource center "ho avy": Climate change resilience building. [Ranobe event post]. Retrieved October 24, 2009, from http://www.350.org/en/node/7313

350.org. (2009i). Fundación Código Aventura por nuestro planeta [Tumero event post]. Retrieved October 24, 2009, from http://www.350.org/es/node/9666

350.org. (2009j). Organizer resources. Retrieved October 24, 2009, from http://www.350.org/resources

350.org. (2009k). Waking-up-call on climate change [Antananarivo event post]. Retrieved October 24, 2009, from http://www.350.org/en/node/8593

350.org. (2009l). Wien: "Human Banner" am Stephansplatz um 14h [Stephansplatz event post]. Retrieved October 24, 2009, from http://www.350.org/en/node/6229

350.org. (2012) Big news: Obama rejects Keystone XL – But we can't stop here. Retrieved September 16, 2012, from http://www.350.org/en/node/28033

Adrian, A. (2009). Civil society in Second Life. *International Review of Law, Computers & Technology*, 23(3), 231–235.

Agenda 21. (1992). Retrieved July 17, 2012, from http://www.un.org/esa/dsd/agenda21/res_agenda21_23.shtml

Agents for Environmental Justice & Scandrett, E. (2003). *Voices from the grassroots*. Edinburgh, Scotland: Friends of the Earth Scotland.

Agyeman, J. (2005). *Sustainable communities and the challenge of environmental justice*. New York, NY: New York University Press.

Aldy, J. E., Barrett, S., & Stavins, R. N. (2003). Thirteen plus one: A comparison of global climate policy architectures. *Climate Policy, 3*, 373–397.

Alexandrova, E. (2011). Metamorphoses of civil society and politics: From Gankos Café to Facebook. In G. Lozanov & O. Spassov (Eds.) *Media and Politics* (pp. 102–117). Sofia, Bulgaria: Konrad Adenauer Stiftung.

Allan, S., Adam, B., & Carter, C. (2000). Introduction: The media politics of environmental risks. In S. Allan, B. Adam, & C. Carter (Eds.), *Environmental Risks and the Media* (pp. 2–26). London, UK: UCL.

Allen, J. (2007, December 13). Interview to The Sidney Morning Herald: 'Avatars' offer virtual alternative at climate summit. Retrieved September 26, 2012 from http://news.smh.com.au/technology/avatars-offer-virtual-alternative-at-climate-summit-20071213-1gxc.html

Altman, R. (1999). *Film/Genre*. London, UK: British Film Institute.

American Psychological Association Task Force. (2009). Report of the American Psychological Association Task Force on the interface between psychology and global climate change. Retrieved August 6, 2010, from http://www.apa.org/science/about/publications/climate-change.aspx

Anderson, A. (1997). *Media, culture and the environment.* London, UK: Routledge.

Andreassen, T., Møller, I. S., & Johansen, N. (2006, October 28). Vi gleder oss til klimaendringer. *Dagsavisen.* Retrieved January 6, 2007, from http://www.dagsavisen.no/

Arnstein, S. R. (1969). A ladder of citizen participation. *Journal of the American Institute of Planners,* 35, 216–224.

Arthus-Bertrand, Y. (2005). *Earth from the air.* London, UK: Thames & Hudson.

Asen, R. (2004). A discourse theory of citizenship. *Quarterly Journal of Speech, 90*(2), 189–211.

Atton, C. (2004). *An alternative Internet: Radical media, politics and creativity.* Edinburgh, Scotland, and New York, NY: Edinburgh University Press & Columbia University Press.

Aufderheide, P. (2006). An Inconvenient Truth. *Cineaste, 32*(1), 50–52.

Aune, M. (2007). Energy comes home. *Energy Policy, 35*(11).

Aune, M., & Berker, T. (2007). Energiforbruk i boliger og yrkesbygg utfordringer og muligheter. In M. Aune & K. H. Sørensen (Eds.), *Mellom klima og komfort: Utfordringer for en bærekraftig energiutvikling* [Between climate and comfort: Challenges for sustainable energy development] (pp. 47–64). Trondheim, Norway: Tapir Akademiske Forlag. (Available only in Norwegian.)

Aune, M., Ryghaug, M., & Godbolt, Å. (2011). Comfort, consciousness and costs: Transitions in Norwegian energy culture, 1991–2010. In *Energy efficiency first: The foundation of a low-carbon society, ECEEE 2011 Summer Study Proceedings.* Stockholm, Sweden: Borg & Co.

Aust, C. F., & Zillmann, D. (1996). Effects of victim exemplification in television news on viewer perception of social issues. *Journalism and Mass Communication Quarterly, 73*, 787–803.

Bäckstrand, K., & Lövbrand E. (2007). Climate governance beyond 2012: Competing discourses of green governmentality, ecological modernization and civic environmentalism. In Pettenger, M. E. (Ed.), *The social construction of climate change: Power, knowledge, norms, discourses* (pp. 123–148). Hampshire, UK: Ashgate.

Balkin, A. (2012). Public Smog/Douala (2009). Retrieved August 3, 2012 from http://tomorrowmorning.net/publicsmog_billboards

Barnouw, E. (1993). *Documentary: A history of the non-fiction film* (2nd rev. ed.). Oxford, UK: Oxford University Press.

Barry, B. (1965). *Political argument.* London, UK: Routledge.

Beattie, K. (2008). *Documentary display: Re-viewing nonfiction film and video.* London, UK: Wallflower Press.

Beck, U. (1992). *Risk society.* London, UK: SAGE.

Beck, U. (1995). *Ecological politics in an age of risk.* Cambridge, UK: Polity.

Beck, U., Giddens, A. & Lash, S. (1996). *Reflexive modernization: Politics, tradition and aesthetics in the modern social order.* Cambridge, UK: Polity Press.

Beers, D. (2009b, March 25). Want one Port Mann bridge, or a light rail metropolis? *The Tyee.* Retrieved March 15, 2010, from http://thetyee.ca/News/2009/03/25/LightRail/

Bellamy Foster, J. (2010, January, 1–18). Why ecological revolution? *Monthly Review, 61*(8), 1–18.

Bellamy, P. H., Loveland, P. J., Bradley, I., Lark, R. M., & Kirk, G. J. D. (2005). Carbon losses from all soils across England and Wales, 1978–2003. *Nature, 437*, 245–248.

Benkler, Y. (2006). *The wealth of networks.* New Haven, CT: Yale University Press.

Bennett, W. L. (2003). New media power: The Internet and global activism. In N. Couldry & J. Currans (Eds.), *Contesting media power* (pp. 17–37). Lanham, MD: Rowman & Littlefield.

Bennett, W. L., & Entman, R. (Eds.). (2001). *Mediated politics in the future of democracy.* Cambridge, UK: Cambridge University Press.

Beresford, C. (2009, December 13). Police, demonstrators flood the streets of Copenhagen, *The Tyee.* Retrieved March 15, 2010, from http://thetyee.ca/Blogs/TheHook/Environment/2009/12/13/FloodStreets/

Berglez, P. (2010). Kritisk diskursanalys. In M. Ekström & L. Larsson (Eds.), *Metoder i kommunikationsvetenskap* (2nd ed.), (pp. 265–288). Lund, Sweden: Studentlitteratur.

Berker, T., Hartman, M., Punie, Y., & Ward, K. (Eds.). (2006). *Domestication of media and technology.* Maidenhead, UK: Open University Press.

Besley, J. C. (2010). Imagining public engagement. *Public Understanding of Science.* Advance online edition, November 1, 2010. doi:10.1177/0963662510379792.

Better Earth in Second Life. (2010). Better Earth in Second Life. Retrieved September 15, 2010, from http://secondlife.com/destination/better-earth

Bickerstaff, K., & Walker, G. (2003). The place(s) of matter: Matter out of place – Public understanding of air pollution. *Progress in Human Geography, 27,* 45–67.

Biggar, J. (2009, December 18). In Copenhagen: The new radicals. *The Tyee.* Retrieved March 15, 2010, from http://thetyee.ca/Opinion/2009/12/18/NewRadicals/

Bird, H., Boykoff, M., Goodman, M., Monbiot, G., & Littler, J. (2009, Winter). The media and climate change. *Soundings, 43,* 47–64.

Boardman, B., Bullock, S., & McLaren, D. (1999). *Equity and the environment: Guidelines for a green and socially just government.* London, UK: Catalyst Pamphlet 5 (with Friends of the Earth).

Bohringer, C., Loschel, A., Moslener, U., & Rutherford, T. F. (2009). EU climate policy up to 2020: An economic impact assessment. *Energy Economics,* 31, S295–S305.

Bora, A., & Hausendorf, H. (n.d.). Communicating citizenship in decision making procedures: Towards an interdisciplinary and cross-cultural perspective [Research proposal]. Retrieved May 12, 2011, from ftp://ftp.cordis.europa.eu/pub/improving/docs/ser_citizen_bora.pdf

Bord, R. J., O'Connor, R. E, & Fisher, A. (2000). In what sense does the public need to understand global climate change? *Public Understanding of Science,* 9, 205–218.

Bord, R. J., O'Connor, R. E., & Fisher, A. (1998). Public perceptions of global warming: United States and international perspectives. *Climate Research,* 11(1), 75–84.

Bortree, D. S., & Seltzer, T. (2009). Dialogic strategies and outcomes: An analysis of environmental advocacy groups' Facebook profiles. *Public Relations Review,* 35, 317–319.

Bosetti, V., Carraro, C., & Tavoni, M. (2008, September). *Delayed participation of developing countries to climate agreements: Should action in the EU and US be postponed?* Fondazione Eni Enrico Mattei Working Papers (Paper 229). Retrieved February 3, 2011, from http://www.econstor.eu/bitstream/10419/53317/1/643917497.pdf

Bostrom, A., & Lashof, D. (2007). Weather or climate change? In S. C. Moser & L. Dilling (Eds.), *Creating a climate for change: Communicating climate change and facilitating social change* (pp. 31–34). New York, NY: Cambridge University Press.

Bostrom, A., Morgan, M. G., Fischhoff, B., & Read, D. (1994). What do people know about climate change? 1: Mental models. *Risk Analysis,* 14(6), 959–970.

Boykoff, M. T., & Boykoff, J. M. (2007). Climate change and journalistic norms: A case-study of US mass-media coverage. *Geoforum,* 38, 1190–1204.

Brackhage, S. (Director). (1981). *The Garden of Earthly Delights* [Motion picture]. USA: Criterion.

Braun, K., & S. Schultz. (2010). "... a certain amount of engineering involved": Constructing the public in participatory governance arrangements. *Public Understanding of Science, 19*(4), 403–419.

Breese, E. B. (2011). Mapping the variety of public spheres. *Communication Theory,* 21, 130–149.

Bricker, K. S., & Kerstetter, D. L. (2000). Level of specialization and place attachment: An exploratory study of whitewater recreation. *Leisure Sciences,* 22, 233–258.

Brook Lyndhurst & Ecometrica. (2011) *Review of the Climate Challenge Fund.* Retrieved June 25, 2012, from http://www.scotland.gov.uk/Publications/2011/06/28142552/0

Brown, E. B. (1967). High resolution aerial photography. *Journal of the Society of Motion Picture and Television Engineers, 76* (7), 100–104.

Brulle, R. J. (2010). From environmental campaigns to advancing the public dialog: Environmental communication for civic engagement. *Environmental Communication: A Journal of Nature and Culture, 4*(1), 82–98.

Bullard, R. D., Johnson, G. S., & Torres, A. O. (2005). Addressing global poverty, pollution, and human rights. In R. D. Bullard (Ed.), *The quest for environmental justice: human rights and the politics of pollution* (pp. 279–297). San Francisco, CA: Sierra Club Books.

Burns, A., & Eltham, B. (2009). Twitter free Iran: An evaluation of Twitter's role in public diplomacy and information operations in Iran's 2009 election crisis. In *Communications policy & research forum 2009,* November 19–20, 2009, University of Technology, Sydney. Retrieved September 15, 2012, from http://vuir.vu.edu.au/15230/

Burris, V. (2005). The future of class analysis: Reflections on "class structure and political ideology." In R. F. Levine (Ed.), *Enriching the sociological imagination: How radical sociology changed the discipline* (pp. 133–163). Boulder, CO: Paradigm.

Burrows, M. (2009a, April 23). BC Green party leader Jane Sterk hopes for MLA seats. *The Georgia Straight.* Retrieved March 15, 2010, from http://www.straight.com/article-215488/greens-hope-mla-seats

Burrows, M. (2009b, April 16). Confronting a climate crisis. *The Georgia Straight*. Retrieved March 15, 2010, from http://www.straight.com/article-214396/confronting-climate-crisis

Burtynsky, E. (n.d.). *Edward Burtynsky*. Retrieved August 4, 2010, from Edward Burtynsky Photographic Works: http://www.edwardburtynsky.com/

Cabecinhas, R., Lázaro, A., & Carvalho, A. (2008). Media uses and social representations of climate change. In A. Carvalho (Ed.), *Communicating climate change: Discourses, mediations and perceptions* (pp. 170–189). Braga, Portugal: Centro de Estudos de Comunicação e Sociedade, Universidade do Minho. E-book available at http://www.lasics.uminho.pt/ojs/index.php/climate_change

Campaign Against Climate Change. (2009). *A million climate jobs*. London, UK: Campaign Against Climate Change.

Carpentier, N. (2011). The concept of participation. *Communication Management Quarterly, 6*(21), 13–36.

Carvalho, A. (2005a). "Governmentality" of climate change and the public sphere. In E. Rodrigues & H. Machado (Orgs.), *Proceedings of the International Seminar Scientific proofs and international justice: The future for scientific standards in global environmental protection and international trade* (pp. 51–69). Braga, Portugal: Núcleo de Estudos em Sociologia, Universidade do Minho.

Carvalho, A. (2005b). Representing the politics of the greenhouse effect: Discursive strategies in the British media. *Critical Discourse Studies, 2*(1), 1–29.

Carvalho, A. (2008, May). *Constructing rights and duties towards climate change: Citizenship and governance in mediated discourses around the world*. Paper presented at the 58th Annual Conference of the International Communication Association, Montreal, Canada.

Carvalho, A. (2010). Media(ted) discourses and climate change: A focus on political subjectivity and (dis)engagement. *WIREs Climate Change, 1*(2), 172–179.

Carvalho, A. (Ed.). (2011). *As alterações climáticas, os media e os cidadãos*. Coimbra, Portugal: Grácio Editor.

References

Carvalho, A., & Pereira, E. (2008). Communicating climate change in Portugal: A critical analysis of journalism and beyond. In A. Carvalho (Ed.), *Communicating climate change: Discourses, mediations and perceptions* (pp. 126–156). Braga, Portugal: Centro de Estudos de Comunicação e Sociedade, Universidade do Minho. E-book available at http://www.lasics.uminho.pt/ojs/index.php/climate_change

Castells, M. (2008). The new public sphere: Global civil society, communication networks, and global governance. *The ANNALS of the American Academy of Political and Social Science, 616*, 78.

Castree, N. (2003). Commodifying what nature? *Progress in Human Geography, 27*(3), 273–297.

Centre for Social Markets. (2012). About Centre for Social Markets. Retrieved July 31, 2012, from http://www.csmworld.org/Table/About-CSM/

Chinese National Development and Reform Commission. (2007). China's national climate change programme. Retrieved July 2, 2012, from www.ccchina.gov.cn/WebSite/CCChina/UpFile/File188.pdf

CNN enters the virtual world of Second Life. (2007). Cable News Network [CNN]. Retrieved October 10, 2008, from http://articles.cnn.com/2007-11-12/tech/second.life.irpt_1_second-life-virtual-world-user-generated-content?_s = PM:TECH

Coal Action Scotland. (2009). *Mainshill: Stories from the woods* [Newsletter]. Anti Copyright. Available from contacts@coalactionscotland.org.uk

Coenen, F. (ed.) (2010). *Public participation and better environmental decisions: The promise and limits of participatory processes for the quality of environmentally related decision-making*. Berlin, Germany: Springer.

Commonwealth Islands in Second Life. (2010). Retrieved August 10, 2010, from http://commonwealthisland.ning.com/

Conners Petersen, L., & Conners, N. (Directors). (2007). *The 11th Hour* [Motion picture]. USA: Warner Independent Pictures.

Corner, A., & A. Randall (2011). Selling climate change? The limitations of social marketing as a strategy for climate change public engagement. *Global Environmental Change,* 21, 1005–1014.

Corruption stalls government attempts to curb CFC trade. (2005). *China Development Brief.* Retrieved November 1, 2007, from http://www.chinadevelopmentbrief.com/node/371

Cosgrove, D., & Fox, W. L. (2010). *Photography and flight.* London, UK: Reaktion Books.

Cotton, C. (2009). *The photograph as contemporary art.* London, UK: Thames & Hudson.

Couldry, N., Livingstone, S., & Markham, T. (2007). *Media consumption and public engagement: Beyond the presumption of attention.* Basingstoke, UK: Palgrave Macmillan.

Cox, R. (2006). *Environmental communication and the public sphere.* London, UK: Sage.

Creasy, S., Gavelin, K., Fisher, H., Holmes, L., & Desai, M. (2007). *Engage for change: The role of public engagement in climate change policy.* Retrieved July 14, 2012, from http://www.sd-commission.org.uk/publications.php?id = 618

Crompton, T. (2010). *Common cause: The case for working with cultural values.* WWF-UK, COIN, CPRE FOE, Oxfam. Retrieved from http://www.wwf.org.uk/what_we_do/campaigning/strategies_for_change

Crowther, J., Galloway, V., & Martin, I. (Eds.). (2005). *Popular education: Engaging the academy.* Leicester, UK: National Institute of Adult Continuing Education.

Crowther, J., Martin, I., & Shaw, M. (Eds.). (1999). Popular education and social movements in Scotland today. Leicester, UK: National Institute of Adult Continuing Education.

Curtice, J., Allaway, A., Robertson, C., Morris, G., Allardice, G., & Robertson, R. (2005). *Public attitudes and environmental justice in Scotland.* Edinburgh, Scotland: Scottish Executive Social Research.

Dahlberg, L. (2001a). Computer mediated communication and the public sphere: A critical analysis. *Journal of Computer Mediated Communication, 7*(1). doi:10.1111/j.1083-6101.2001.tb00137.x

Dahlberg, L. (2001b). The Internet and democratic discourse: Exploring the prospects of online deliberative forums extending the public sphere. *Information, Communication & Society, 44*, 615–633.

Dahlberg, L., & Siapera, E. (Eds.). (2007) *Radical democracy and the Internet: Exploring theory and practice.* London, UK: Palgrave.

Dahlgren, P. (1995). *Television and the public sphere.* London, UK: Sage.

Dahlgren, P. (2005). The Internet, public spheres, and political communication: Dispersion and deliberation. *Political Communication, 22*(2), 147–162.

Dahlgren, P. (2006). Doing citizenship: The cultural origins of civic agency in the public sphere. *European Journal of Cultural Studies, 9*(3), 267–286.

Dahlgren, P. (2009). *Media and Political Engagement.* Cambridge, UK: Cambridge University Press.

Dahlgren, P. (2011). Parameters of online participation: Conceptualising civic contingencies. *Communication Management Quarterly, 6*(21), 87–110.

Danielsson, J. (2007). Svensk ambassad i cyberrymden. Retrieved October 10, 2008, from http://www.gp.se/ekonomi/1.155674-svensk-ambassad-i-cyberrymden

David, L. (Producer), & Guggenheim, D. (Director). (2006). *An Inconvenient Truth* [Motion picture]. USA: Paramount Classics.

David, L., & Gordon, C. (2007). *The down-to-earth guide to global warming.* New York, NY: Orchard Books, Scholastic Inc.

Davis, J. E. (2002). Narrative and social movements: The power of stories. In J. E. Davis (Ed.), *Stories of change: Narrative and social movements* (pp. 3–30). New York, NY: State University of New York Press.

Dean, M. (1999). *Governmentality.* London, UK: Sage.

Delgado, A., Kjølberg, K. L., & Wickson, F. (2010, May 11). Public engagement coming of age: From theory to practice in STS encounters with nanotechnology. *Public Understanding of Science*. Advance online publication. doi:10.1177/0963662510363054.

DeLuca, K. M (1999). *Image politics: The new rhetoric of environmental activism*. New York, NY: Guilford Press.

Dembicki, G. (2009, April 23). Premier has rosy view of BC's carbon targets. *The Tyee*. Retrieved March 15, 2010, from http://thetyee.ca/Blogs/TheHook/BC-Politics/2009/04/23/emissions-targets-experts-campbell/

Demeritt, D. (2001). The construction of global warming and the politics of science. *Annals of the Association of American Geographers, 91*(2), 307–337.

Department of Energy and Climate Change (UK). (2009). 5 NC: The UK's fifth national communication under the United Nations Framework Convention on Climate Change. London, UK: Department of Energy and Climate Change.

Diani, M. (2000). Social movement networks: Virtual and real. *Information, Communication & Society, 3*(3), 386–401.

Dierkes, M., & von Grote, C. (Eds.). (2000). *Between understanding and trust: The public, science and technology*. Amsterdam, The Netherlands: Harwood.

Dietz, T., & P. C. Stern (Eds.). (2008). *Public participation in environmental assessment and decision making*. Washington, DC: National Academies Press.

Dow, K., & Downing, E. (2006). *The atlas of climate change: Mapping the world's greatest challenge*. Berkeley: University of California Press.

Doyle, J. (2007). Picturing the Clima(c)tic: Greenpeace and the representational politics of climate change communication. *Science as Culture, 16*(2), 129–150.

Doyle, J. (2009). Seeing the climate? The problematic status of visual evidence in climate change campaigning. In S. I. Dobrin & S. Morey (Eds.), *Ecosee: Image, Rhetoric, Nature* (pp. 279–298). New York, NY: State University of New York Press.

Dryzek, J. S. (2000). *Deliberative democracy and beyond: Liberals, critics, contestations.* Oxford, UK: Oxford University Press.

Dryzek, J. S. (1994). *Discursive democracy: Politics, policy, and political science.* Cambridge, UK: Cambridge University Press.

Dryzek, J. S. (2005). *The Politics of the earth: Environmental discourses* (2nd ed.). Oxford, UK: Oxford University Press.

Dryzek, J. S., Downs, D., Hernes, H.-K., & Schlosberg, D. (2003). *Green states and social movements: Environmentalism in the US, UK, Germany and Norway.* Oxford, UK: Oxford University Press.

Dunaway, F. (2005). *Natural visions: The power of images in American environmental reform.* Chicago: University of Chicago Press.

Dunion, K. (2003). *Troublemakers: The struggle for environmental justice in Scotland.* Edinburgh, Scotland: Edinburgh University Press.

Dunion, K., & Scandrett, E. (2003). The campaign for environmental justice in Scotland as a response to poverty in a northern nation. In J. Agyeman, R. D. Bullard, & B. Evans (Eds.), *Just sustainabilities: Development in an unequal world* (pp. 311–322). London, UK: Earthscan.

Dunlap, R. E., & McCright, R. E. (2010). Climate change denial: Sources, actors, strategies. In C. Lever-Tracy (Ed.), *Routledge handbook of climate change and society* (pp. 240–260). Oxon, UK: Routledge.

Dyer, G. (2009a, April 7). Gwynne Dyer: Barack Obama's emission cuts are pragmatic suicide. *The Georgia Straight.* Retrieved March 15, 2010, from http://www.straight.com/article-213077/gwynne-dyer-barack-obamas-emission-cuts-are-pragmatic-suicide

Dyer, G. (2009b, December 21). Gwynne Dyer: The aftermath of Copenhagen. *The Georgia Straight.* Retrieved March 15, 2010, from http://www.straight.com/article-276143/vancouver/gwynne-dyer-aftermath-copenhagen

ecoAmerica, Western Strategies, & Lake Research Partners. (2009). *Summary report: Climate and energy truths: Our common future.* Retrieved October 4, 2010, from http://ecoamerica.net/press/media/090520/truths

Ehrenreich, B., & Ehrenreich, J. (1977). The professional-middle class. *Radical America, 11*(2), 7–31.

Endres, D., Sprain, L., & Peterson, T. R. (2009). *Social movement to address climate change: Local steps for global action.* Amherst, NY: Cambria Press.

Entman, R. M. (1993). Framing: Toward clarification of a fractured paradigm. *Journal of Communication, 43*(4), 51–58.

Etopia Eco Village in Second Life. (2010). Ectopia Eco Village in Second Life. Retrieved September 14, 2010, from http://world.secondlife.com/group/bfb5e591-ec12-ced0-9a30-5be0c57be867

Evelyn, J. (1661). *Fumifugium.* London, UK: W. Godbid for Gabriel Bedel & Thomas Collins.

Faber, D. R., & McCarthy, D. (2003). Neo-liberalism, globalization and the struggle for ecological democracy: Linking sustainability and environmental justice. In J. Agyeman, R. D. Bullard, & B. Evans (Eds.), *Just sustainabilities: Development in an unequal world* (pp. 39–63). London, UK: Earthscan.

Facebook user. (2010). Facebook comment in the group Greenpeace International. Retrieved September 25, 2010, from http://www.facebook.com/search/?init = quick&q = greenpeace%20in&ref = ts#!/topic.php?uid = 7297163299&topic = 14928

Fairclough, N. (2001). *Language and Power* (2nd ed.). Toronto, Canada: Pearson Education.

Fairclough, N. (2003). *Analysing discourse: Textual analysis for social research.* London, UK: Routledge.

Fairclough, N. (2006). *Language and globalization.* London, UK: Routledge.

Federal Communications Commission [FCC]. (2007). Comission opens inquiry on competitive bidding process for report to Congress. Retrieved October 21, 2007, from http://wireless.fcc.gov/auctions/general/releases/fc970232.txt

Feldpausch-Parker, A. M. (2010). *Communicating carbon capture and storage technologies: Opportunities and constraints across media.*

(Unpublished PhD thesis). Texas A&M University, College Station, TX.

Felt, U., & Fochler, M. (2008). The bottom-up meanings of the concept of public participation in science and technology. *Science and Public Policy, 35*(7), 489–499.

Fischer, F. (2000). *Citizens, experts, and the environment: The politics of local knowledge*. Durham, NC: Duke University Press.

Flannery, T. (2005). *The weather makers*. New York, NY: Grove Press.

Fletcher, A. L. (2009). Clearing the air: The contribution of frame analysis to understanding climate policy in the United States. *Environmental Politics, 18*(5), 800–816.

Flowerdew, J. (2008). Critical discourse analysis and strategies of resistance. In V. K. Bhatia, J. Flowerdew, & R. H. Jones (Eds.), *Advances in Discourse Studies* (pp. 195–210). London, UK: Routledge.

Fornäs, J., Klein, K., Ladendorf, M., Sundén, J., & Sveningsson, M. (Eds.) (2002). *Digital borderlands: Cultural studies of identity and interactivity on the Internet*. New York, NY: Peter Lang.

Foucault, M. (1991). Governmentality. In G. Burchell, C. Gordon, & P. Miller (Eds.), *The Foucault effect* (pp. 87–104). Chicago, IL: University of Chicago Press.

Foundas, S. (2007, March 5). Everything's Cool [Film review]. *Variety*, p. 33.

Foust, C. R., & Murphy, W. O. (2009). Revealing and reframing apocalyptic tragedy in global warming discourse. *Environmental Communication: A Journal of Nature and Culture, 3*(2), 151–167.

Fraser, N. (1992). Rethinking the public sphere: A contribution to the critique of actually existing democracy. In C. Calhoun (Ed.), *Habermas and the public sphere* (pp. 109–142). Cambridge, MA: MIT Press.

Freire, P. (1972). *Pedagogy of the oppressed*. London, UK: Penguin.

Friends of the Earth International. (2010). Climate justice briefs. Retrieved September 18, 2011, from http://www.foe.co.uk/community/campaigns/climate/2010_cancun_briefs_32027.html

Friends of the Earth U.S. (2009). *Subprime carbon? Rethinking the world's largest new derivatives market.* Washington, DC: Friends of the Earth. Retrieved September 10, 2011, from www.foe.org/sites/default/files/SubprimeCarbon_2pager.pdf

From PUS to PEST. (2002). *Science,* 298 (5591), 49b.

Futerra Communications. (2009). Sell the sizzle: The new climate message. Retrieved October 15, 2010, from http://www.futerra.co.uk/downloads/Sellthesizzle.pdf

Gastil, J. J., & Levine, P. (Eds.) (2005). *The deliberative democracy handbook: Strategies for effective civic engagement in the 21st century.* San Francisco, CA: Jossey-Bass.

Gelinas, N. (2007). A carbon tax would be cleaner. *The Wall Street Journal.* Retrieved August 23, 2007, from http://www.manhattan-institute.org/html/_wsj-a_carbon_tax_would_be_cleaner.htm

The Georgia Straight. (2010). Association of Alternative Newsweeklies. Retrieved October 22, 2010, from http://www.altweeklies.com/aan/the-georgia-straight/Company?oid = 4626

Giddens, A. (1984). *The constitution of society: Outline of the theory of structuration.* Berkeley: University of California Press.

Giddens, A. (2009). *The politics of climate change.* Cambridge, UK: Polity Press.

Gillis, J. (2012, August 6). Study finds more of Earth is hotter and says global warming is at work. *The New York Times.* Retrieved August 6, 2012, from http://www.nytimes.com/2012/08/07/ science/earth/extreme-heat-is-covering-more-of-the-earth-a-study-says.html?emc=tnt&tntemail0=y

Global Warming?? (Facebook group). (2010). About. Retrieved September 15, 2010, from http://www.facebook.com/#!/groups/2374404880/

Gold, D. B., & Helfand, J. (Directors). (2002). *Blue Vinyl* [Motion Picture]. USA: Bullfrog Films.

Gold, D. B., & Helfand, J. (Directors). (2007). *Everything's Cool* [Motion Picture]. USA: City Lights Pictures, Toxic Comedy Pictures, Lupine Films, the Kendeda Sustainability Fund.

Gore, A. (2006). *Earth in the balance.* New York, NY: Rodale Books.

Goulder, L. H., & Stavins, R. N. (2011). Critical issues in national climate policy design: Challenges from state-federal interactions in US climate change policy. *American Economic Review: Papers & Proceedings, 101*(3), 253–257.

Gray, J. (2010, August 2). Unnatural selection [Review of M. Ridley, *The rational optimist: How* prosperity evolves]. *New Statesman.* Retrieved June 25, 2012, from http://www.newstatesman.com/books/2010/08/ridley-climate-evolution-ideas

Greenpeace International. (2010a). Issues we work on/Climate change. Retrieved September 25, 2010, from http://www.greenpeace.org/international/

Greenpeace International. (2010b). Issues we work on/Climate change/Solutions. Retrieved September 25, 2010, from http://www.greenpeace.org/international/

Greenpeace's Second Life: Is the media giant missing the virtual boat?. (2008). Sunshine Communications. Retrieved October 25, 2008, from http://sunshinecommunications.ca/html/modules.php?op = modload&name = News&file = article&sid = 109

Gripsrud, J. (2009). Digitising the public sphere: Two key issues. *Javnost: The Public, 16*(1), 5–16.

Griskevicius, V., Cialdini, R. B., & Goldstein, N. J. (2008). Social norms: An underemployed lever for managing climate change. *International Journal of Sustainability Communication, 3*, 5–13.

Gunster, S. (2011). Covering Copenhagen: Climate politics in B.C. media. *Canadian Journal of Communication, 36*(3), 477–502.

Gurevitch, M., Coleman, S., & Blumler, J. G. (2009). Political communication: Old and new media relationships. *The ANNALS of the American Academy of Political and Social Science, 625*, 164.

Gutkind, E. A. (1952). *Our world from the air: An international survey of man and his environment.* London, UK: Chatto & Windus.

Habermas, J. (1989). *The structural transformation of the public sphere.* Cambridge, MA: MIT Press.

Habermas, J. (1996). *Between facts and norms: Contributions to a discourse theory of law and democracy.* Cambridge, MA: MIT Press.

Hackett, B., & Zhao, Y. (1998). *Sustaining democracy? Journalism and the politics of objectivity.* Toronto, Canada: Garamond Press.

Haddon, L. (2006). The contribution of domestication research to in-home computing and media consumption. *The Information Society, 22*(4), 195–203.

Halliday, M. (1985). *Introduction to functional grammar.* London, UK: Edward Arnold.

Hannigan, J. (2006). *Environmental sociology* (2nd ed.). London, UK: Routledge.

Hansen, A. (2010). *Environment, media and communication.* London, UK: Routledge.

Hardin, G. (1968). The tragedy of the commons. *Science, 162,* 1245.

Harvey, D. (1996). *Justice, nature and the geography of difference.* Oxford, UK: Blackwell.

Harvey, D. (1999). The environment of justice. In F. Fischer, & M. A. Hajer (Eds.), *Living with nature: Environmental politics as cultural discourse* (pp. 153–185). Oxford, UK: Oxford University Press.

Harvey, D. (2005). *A brief history of neoliberalism.* Oxford, UK: Oxford University Press.

Harvey, D. (2006). *Spaces of global capitalism: Towards a theory of uneven geographical development.* London, UK: Verso.

Heller, C. (2004). Risky science and savour-faire: Peasant expertise in the French debate over genetically modified crops. In M. E. Lien & B. Nerlich (Eds.), *The politics of food* (pp. 81–99). Oxford, UK: Berg.

Herzog, W. (Director) (1992). *Lessons of Darkness* [Motion Picture]. France, UK, Germany: Canal+, Première, Werner Herzog Productions.

Herzog, W. (2002). *Herzog on Herzog*. (P. Cronin, Ed.). London, UK: Faber & Faber.

Hiskes, J. (2009, December 7). In a frenzied Copenhagen, crucial climate talks begin. *The Tyee*. Retrieved March 15, 2010, from http://thetyee.ca/Blogs/TheHook/Environment/2009/12/07/FrenziedCopenhagen/

Holden, S. (2007, November 23). Everything's Cool [Film review]. *The New York Times*, p. 13.

Hope, A., & Timmel, S. (1984). *Training for transformation: A handbook for community workers*. Gweru, Zimbabwe: Mambo Press.

Hopkins, R. (2008). *The Transition handbook: From oil dependency to local resilience*. Totnes, UK: Green Books.

Hopkins, R. (2011). *The Transition companion: Making your community more resilient in uncertain times*. Totnes, UK: Green Books.

Hornmoen, H. (2009). "What researchers now can tell us": Representing scientific uncertainty in journalism. *Observatorio, 3*(4), 74–100.

House of Lords Select Committee on Science and Technology. (2000). *Science and society: Third report of the session 1999–2000*. London, UK: HM Stationery Office.

Huang, S. T., Kamel Boulos, M. N., & Dellavalle, R. P. (2008). Scientific discourse 2.0: Will your next poster session be in Second Life? *EMBO reports, 9*(6), 496–499.

Hui, S. (2009, April 19). BC Liberal, NDP election platforms ignore peak oil, but Greens note threat. *The Georgia Straight*. Retrieved March 15, 2010, from http://www.straight.com/article-215121/bc-liberal-ndp-election-platforms-ignore-peak-oil-greens-note-threat

Huitema, D., Cornelisse, C., & Ottow, B. (2010). Is the jury still out? Toward greater insight in policy learning in participatory decision processes—the case of Dutch citizens' juries on water management in the Rhine Basin. *Ecology and Society 15*(1), 16. Retrieved June 23, 2012, from http://www.ecologyandsociety.org/vol15/iss1/art16/

Hulme, M. (2007). Newspaper scare headlines can be counter-productive. *Nature,* 445, 818.

Hulme, M. (2009). *Why we disagree about climate change: understanding controversy, inaction and opportunity.* Cambridge, UK: Cambridge University Press.

Hyde, M. J. (1990). Experts, rhetoric, and the dilemmas of medical technology: Investigating a problem of progressive ideology. In M. J. Medhurst, A. Gonzalez, & T. R. Peterson (Eds.), *Communication and the culture of technology* (pp. 115–136). Pullman: Washington State University Press.

Intergovernmental Panel on Climate Change [IPCC]. (2007a). Summary for Policymakers. In S. Solomon, D. Qin, M. Manning, Z. Chen, M. Marquis, K.B. Averyt, M.Tignor and H.L. Miller (eds.) *Climate change 2007: The physical science basis. Contribution of working group I to the fourth assessment report of the Intergovernmental Panel on Climate Change.* Cambridge, UK and New York, NY: Cambridge University Press.

Intergovernmental Panel on Climate Change [IPCC]. (2007b). *Climate Change 2007: Synthesis Report.* Retrieved January 19, 2010, from www.ipcc.ch/pdf/assessment-report/ar4/syr/ar4_syr.pdf

Intergovernmental Panel on Climate Change [IPCC]. (2007c). *Contribution of Working Group II to the Fourth Assessment Report of the Intergovernmental Panel on Climate Change.* M.L. Parry, O. F. Canziani, J. P. Palutikof, P. J. van der Linden, & C. E. Hanson (Eds.). Cambridge, UK: Cambridge University Press.

International Climate Justice Network, Bali principles of climate justice. (August 28, 2002). *India Resource Center.* Retrieved September 12, 2012, from http://www.indiaresource.org/issues/energycc/2003/baliprinciples.html

Irwin, A., & Michaels, M. (2003). *Science, social theory and public knowledge.* Milton Keynes, UK: Open University Press.

Ivie, R. L. (2001). Hierarchies of equality: Positive peace in a democratic idiom. In G. Cheney, S. May, & D. Munshi (Eds.), *The handbook on communication ethics* (pp. 374–386). New York, NY: Routledge.

Ivie, R. L. (2004). Prologue to democratic dissent in America. *Javnost, The Public 11*(2), 19–35.

Jackson, N., & Lilleker, D. (2009). Building an architecture of participation? Political parties and Web 2.0 in Britain. *Journal of Information Technology and Politics, 6*(3/4), 232–250.

Jackson, T. (2009). *Prosperity without growth: The transition to a sustainable economy*. Sustainable Development Commission. Retrieved September 3, 2010, from www.sd-commission.org.uk/publications/downloads/prosperity_without_growth_report.pdf.

Jäger, S., & Maier, F. (2009). Theoretical and methodological aspects of Foucauldian critical discourse analysis and dispositive analysis. In R. Wodak & M. Meyer (Eds.), *Methods of critical discourse analysis* (pp. 34–61). London, UK: Sage.

Jamison, A. (2001). *The making of green knowledge: Environmental politics and cultural transformation*. Cambridge, UK: Cambridge University Press.

Jasanoff, S. (2004). Heaven and earth: The politics of environmental images. In S. Jasanoff & M. Martello (Eds.), *Earthly politics: Local and global in environmental governance* (pp. 31–55). Cambridge, MA & London, UK: MIT Press.

Jasanoff, S. (2005). *Designs on nature: Science and democracy in Europe and the United States*. Princeton, NJ: Princeton University Press.

Joas, M., Kern, K., & Sandberg, S. (2007). Actors and arenas in hybrid networks: Implications for environmental policymaking in the Baltic Sea region. *Ambio: A Journal of the Human Environment, 36*(2), 237–242.

Johansen, N. (2006, November 10). Blåser i Bjørnøys klimakrav. *Dagsavisen*. Retrieved February 10, 2007, from http://www.dagsavisen.no/

Join the Fight to Stop Global Warming! (Facebook group). (2010). About. Retrieved September 15, 2010, from http://www.facebook.com/#!/groups/20009925629/

Jönsson, A. M., & Örnebring, H. (2011). User-generated content and the news: Empowerment of citizens or interactive illusion? *Journalism Practice, 5*(2), 127–144.

Kahn, R., & Kellner, D. (2005). Oppositional politics and the Internet: A critical/reconstructive approach. *Cultural Politics, 1*(1), 75–100.

Kane, L. (2001). *Popular education and social change in Latin America.* London, UK: Latin America Bureau.

Karlstrøm, H. (2010). *Survey: Den deregulerte forbruker* [Report]. Department of Interdisciplinary Studies of Culture, Norwegian University of Science and Technology, Trondheim, Norway.

Kellner, D. (2010). *Cinema Wars: Hollywood film and politics in the Bush-Cheney era.* Malden, UK: Wiley-Blackwell.

Kempton, W., Boster, J. S., & Hartley, J. A. (1995). *Environmental values in American culture.* Cambridge, MA: MIT Press.

Kettlewell, D. (2009, April 30). Damian Kettlewell: A green vision for BC and Vancouver–False Creek. *The Georgia Straight.* Retrieved March 15, 2010, from http://www.straight.com/article-217200/damian-kettlewell-green-vision-bc-and-vancouverfalse-creek

Kiehl, J. (2011). Lessons from Earth's past. *Science, 331,* 158–159.

Kimmett, C. (2009a, December 8). Gateway at odds with BC's carbon goals, say protesters. *The Tyee.* Retrieved March 15, 2010, from http://thetyee.ca/Blogs/TheHook/Environment/2009/12/08/GatewayCarbon/

Kimmett, C. (2009b, December 9). Canada vilified in Copenhagen. *The Tyee.* Retrieved March 15, 2010, from http://thetyee.ca/Blogs/TheHook/Environment/2009/12/09/Canadavilified/

Kingsnorth, P. (2009). *Uncivilisation: The Dark Mountain manifesto.* Retrieved September 3, 2010, from www.dark-mountain.net/wordpress/dark-mountain.net/wordpress/wp-content/uploads//uncivilisation-dark-mountain-manifesto.pdf.

Kitzinger, J., & Reilly, J. (1997). The rise and fall of risk reporting: Media coverage of human genetics research, "False Memory Syndrome," and "Mad Cow Disease." *European Journal of Communication, 12*, 319–50.

Klein, N. (2009, November 13). Naomi Klein: Copenhagen climate change summit not merely a Seattle WTO do-over. Retrieved March 15, 2010, from http://www.straight.com/article-270383/vancouver/naomi-klein-copenhagen-climate-change-summit-not-merely-seattle-wto-doover

Klein, N. (2007) *The shock doctrine: the rise of disaster capitalism*. Toronto: Alfred A. Knopf Canada.

Kleinman, D., Delborne, J., & Anderson, A. (2011). Engaging citizens: The high cost of citizen participation in high technology. *Public Understanding of Science, 20*(2), 221–240.

Klinke, O., & Renn, A. (2002). A new approach to risk evaluation and management: Risk-based, precaution-based and discourse-based strategies. *Risk Analysis, 22*, 1071–1094.

Klotz, R. (2001). Internet politics: A survey of practices. In R. P. Hart & D. R. Shaw (Eds.), *Communication in US elections: New agendas* (pp. 185–201). Lanham, MD: Rowman & Littlefield.

Krasner, S. (1982). Structural causes and regime consequences: Regimes as intervening variables. *International Organization, 36*, 185–205.

Kyoto Protocol. (1997). Retrieved July 16, 2012, from http://unfccc.int/kyoto_protocol/items/2830.php/

Laclau, E., & Mouffe, C. (1985). *Hegemony and socialist strategy*. London, UK: Verso.

Laclau, E., & Mouffe, C. (2001). *Hegemony and socialist strategy: Towards a radical democratic politics*. London, UK: Verso.

Lahsen, M. (2007). Trust through participation? Problems of knowledge in climate decision making. In M. E. Pettenger (Ed.), *The social construction of climate change: Power, knowledge, norms, discourses* (pp. 173–196). Hampshire, UK: Ashgate.

Laird, F. N. (1990). Technocracy revisited: Knowledge, power and the crisis in energy decision making. *Industrial Crisis Quarterly, 4*, 49–61.

Lakoff, G. (2010). Why it matters how we frame the environment. *Environmental Communication: A Journal of Nature and Culture,* 4(1), 70–81.

Lassen, I., Horsbøl, A., Bonnen, K., & Pedersen, A. G. J. (2011). Climate change discourses and citizen participation: A case study of the discursive construction of citizenship in two public events. *Environmental Communication: A Journal of Nature and Culture,* 5(4), 411–427.

Lawson, N. (2008). *An appeal to reason: A cool look at global warming.* London, UK: Overlook Duckworth.

Lefebvre, H. (1974/1991). *The production of space.* D. Nicholson-Smith (Trans.). Oxford, UK: Blackwell.

Leiserowitz, A. (2005). American risk perceptions: Is climate change dangerous? *Risk Analysis,* 25(6), 1433–1442.

Leiserowitz, A. A. (2007, October). *Assessing the public impact: The Day After Tomorrow, An Inconvenient Truth, and LiveEarth.* Presentation at the Behavior, Energy, and Climate Change Conference, Sacramento, CA.

Leiserowitz, A., & Fernandez, L. O. (2008). Toward a new consciousness: Values to sustain human and natural communities. *Environment: Science and Policy for Sustainable Development,* 50(5), 62–69.

Lewis, J., Wahl-Jorgensen, K., & Inthorn, S. (2004). Images of citizenship on television news: Constructing a passive public. *Journalism Studies,* 5(2), 153–164.

Lie, M., & Sørensen, K. H. (Eds.) (1996). *Making technology our own? Domesticating technology into everyday life.* Oslo, Norway: Scandinavian University Press.

Lin, A. (2007). Virtual consumption: A Second Life for Earth? [Unpublished paper]. Retrieved from http://works.bepress.com/albert_lin/1/

Linden, T. (2010, January 19). Web log comment: '2009 End of Year Second Life Economy Wrap up (including Q4 Economy in Detail)', Second Life, Community, Blogs. Retrieved September 14, 2010, from http://blogs.secondlife.com/community/features/blog/2010/01

/19/2009-end-of-year-second-life-economy-wrap-up-including-q4-economy-in-detail

Lippard, L., & Chandler, J. (1968, February). The dematerialization of art. *Art International 12*(2), 31–36.

Lippard, L. (1973). *Six years: The dematerialization of the art object from 1966 to 1972.* New York, NY: Praeger.

Lohman, L. (Ed.). (2006). *Carbon trading: A critical conversation on climate change, privatisation and power.* Special issue of *Development Dialogue*, 48.

Lohmann, L. (2008) Carbon trading, climate justice and the production of ignorance: 10 examples. *Development*, 51, 359–365.

Lomborg, B. (2001). *The skeptical environmentalist.* Cambridge, UK: Cambridge University Press.

López, A. (2010). Defusing the cannon/canon: An organic media approach to environmental communication, *Environmental Communication, 4*(1), 99–108.

Lorenzoni, I., & Pidgeon, N. (2006). Public views on climate change: European and USA perspectives. *Climatic Change*, 77, 73–95.

Lorenzoni, I., Nicholson-Cole, S., & Whitmarsh, L. (2007). Barriers perceived to engaging with climate change among the UK public and their policy implications. *Global Environmental Change*, 17, 445–459.

Low, S. (2000). *On the plaza: The politics of public space and culture.* Austin, TX: University of Texas Press.

Lowe, T., Brown, K., Dessai, S., Doria, M. F., Haynes, K., & Vincent, K. (2006). Does tomorrow ever come? Disaster narrative and public perception of climate change. *Public Understanding of Science*, 15, 435–457.

Luke, T. (1995). Sustainable development as a power/knowledge system: The problem of "governmentality." In F. Fischer & M. Black (Eds.), *Greening environmental policy: The politics of a sustainable future* (pp. 21–32). London, UK: Paul Chapman.

Luke, T. W. (2008). The politics of true convenience or inconvenient truth: Struggles over how to sustain capitalism, democracy, and ecology in the 21st century. *Environment and Planning A, 40,* 1811–1824.

Lupick, T. (2009, April 30). BC Greens' energy plan depend less on run of river. *The Georgia Straight.* Retrieved March 15, 2010, from http://www.straight.com/article-217175/bc-greens-energy-plans-depend-less-run-river

M'Gonigle, M. (2009a, April 20). Tapping our wild rivers can't fix climate change. *The Tyee.* Retrieved March 15, 2010, from http://thetyee.ca/Views/2009/04/20/PowerDown/

M'Gonigle, M. (2009b, December 6). Against Copenhagen. *The Tyee.* Retrieved March 15, 2010, from http://thetyee.ca/Opinion/2009/12/06/CopenhagenContradictions/

M'Gonigle, M. (2009c, December 8). The elephants of doom in Copenhagen. *The Tyee.* Retrieved March 15, 2010, from http://thetyee.ca/Opinion/2009/12/08/ElephantsOfDoom/

M'Gonigle, M., & Anderson, B. (2009, April 30). Beyond the carbon tax. *The Tyee.* Retrieved March 15, 2010, from http://thetyee.ca/Views/2009/04/30/CarbonTax/

MacDonald, K. A. (2007, January 18). Special Report 1. *Daily Variety,* p. A9.

MacDonald, S. (2001). *The garden in the machine: A field guide to independent films about place.* Berkeley: University of California Press.

Maibach, E., Roser-Renouf, C., & Leiserowitz, A. (2009). Global warming's six Americas: An audience segmentation. New Haven, CT: Yale University & George Mason University.

Mansell, R. (2004). Political economy, power and new media. *New Media & Society, 6*(1), 96–105.

Martin, E. (1994). *Flexible bodies: Tracking immunity in American culture from the days of polio to the age of AIDS.* Boston, MA: Beacon Press.

Martinez-Alier, J. (2002). *Environmentalism of the poor: A study of ecological conflicts and valuation.* Cheltenham, UK: Edward Elgar.

Mazzoleni, G., & Schulz, W. (1999). Mediatization of politics: A challenge for democracy? *Political Communication, 16*(3), 247–261.

McCormick, J. (1991). *Reclaiming paradise: The global environmental movement.* Bloomington: Indiana University Press.

McCright, A. M., & Dunlap, R. E. (2000). Challenging global warming as a social problem: An analysis of the conservative movement's counterclaims. *Social Problems, 47*(4), 499–522.

McKibbin, W. J., & Wilcoxen, P. J. (2002). The role of economics in climate change policy. *Journal of Economic Perspectives, 16*(2), 107–129.

McLaren, D. (2003). Environmental space, equity and the ecological debt. In J. Agyeman, R.D. Bullard, & B. Evans (Eds.), *Just sustainabilities: Development in an unequal world* (pp. 19–37). London, UK: Earthscan.

Meek, T. (2007). The 11th Hour [Film review]. *Cineaste, 33*(1), 58–59.

Methmann, C. P. (2010). "Climate protection" as empty signifier: A discourse theoretical perspective on climate mainstreaming in world politics. *Millennium: Journal of International Studies, 39*(2), 345–372.

Michael, M. (2002). Comprehension, apprehension, prehension: Heterogeneity and the public understanding of science. *Science, Technology, & Human Values, 27*(3), 357–378.

Miles, M. (2010). Representing nature: Art and climate change. *Cultural Geographies, 17*(1), 19–35.

Ministério da Economia & Ministério das Cidades, Ordenamento do Território e do Ambiente (2004). Plano Nacional de Atribuição de Licenças de Emissão (PNALE) 2005–2007: Relatório da consulta pública. Retrieved July 22, 2012, from http://www.apambiente.pt/_zdata/DPAAC/CELE/PNALE_relatorio_cons_publ_Maio2004.pdf

Ministry of Environment and Forests (India). (2012) Second national communication to the United Nations Framework Convention on Climate Change. New Delhi, India: Ministry of Environment and Forests.

Ministry of Natural Resources and Environment (Tuvalu). (1999). Initial national communication to the United Nations Framework Convention on Climate Change. Tuvalu: Ministry of Natural Resources and Environment.

Minster, M. (2010). The rhetoric of ascent in An Inconvenient Truth and Everything's Cool. In P. Willoquet-Maricondi (Ed.), *Framing the world: Explorations in ecocriticism and film* (pp. 25–42). Charlottesville: University of Virginia Press.

Miraftab, F. (2004). Invited and invented spaces of participation: Neoliberal citizenship and feminists' expanded notion of politics. *Wagadu*, 1, 1–7.

Moore, R. L., & Scott, D. (2002). Place attachment and context: Comparing a park and a trail within. *Forest Science, 49*(6), 877–884.

Morgan, D. L. (1997). *Focus groups as qualitative research* (2nd ed.). Thousand Oaks, CA: Sage.

Mormont, M., & Dasnoy, C. (1995). Source strategies and the mediatization of climate change. *Media, Culture & Society*, 17, 49–64.

Moser, S. (2009). Communicating climate change and motivating civic action: Renewing, activating, and building democracies. In H. Selin & S. D. VanDeever (Eds.), *Changing Climates in North American Politics: Institutions, Policymaking, and Multilevel Governance* (pp. 283–302). Cambridge, MA: MIT Press.

Moser, S. (2010). Communicating climate change: History, challenges, processes, and future directions. *WIREs Climate Change*, 1, 31–53.

Moser, S. C. & Dilling, L. (2004). Making climate hot. *Environment, 46*(10), 32–46.

Moser, S. C., & Dilling, L. (Eds.) (2007). *Creating a climate for change: Communicating climate change and facilitating social change.* Cambridge, UK: Cambridge University Press.

Mouffe, C. (1993/2005). *The return of the political.* London, UK: Verso.

Mouffe, C. (2000). *The democratic paradox.* London, UK: Verso.

Mouffe, C. (2005). *On the political.* London, UK: Routledge.

Mouffe, C. (2006). Which public space for critical artistic practices? In T. Byrne (Ed.), *Cork caucus: On art, possibility, and democracy* (pp. 149–171). Cork, Ireland: National Sculpture Factory & Revolver Press.

Mouffe, C. (2007). Artistic activism and agonistic spaces. *Art and Research: A Journal of Ideas, Contexts and Methods, 1*(2), 6. Retrieved

September 12, 2011, from http://www.artandresearch.org.uk/v1n2/mouffe.html

National Park Service. (2007). *Climate change.* [Brochure]. United States: National Park Service.

National Park Service. (2008a). National Park Service Public Use Statistics Office. Retrieved August 12, 2008, from http://www.nature.nps.gov/stats/viewReport.cfm

National Park Service. (2008b). *Climate Change in Rocky Mountain National Park: Preservation in the Face of Uncertainty.* Retrieved August 12, 2008, from http://www.nps.gov/romo/parkmgmt/upload/climate_change_rocky_mountain2.pdf

National Park Service. (2008c). Glacier Visitor Guide [Newsletter]. United States: National Park Service.

National Park Service. (2008d). *Yellowstone resources and issues: An annual compendium of information about Yellowstone National Park.* Mammoth Hot Springs, WY: Yellowstone Association.

National Park Service. (2008e). *Climate change in national parks* [Brochure]. United States: National Park Service.

National Park Service. (2008f). Report released on climate change issues in Rocky Mountain National Park. Retrieved from http://www.nps.gov/romo/parknews/pr_climate_change_rept.htm

National Park Service. (2008g). *Rocky Mountain National Park Spring Newsletter.* United States: National Park Service.

National Park Service. (2012a). National Park Service Public Use Statistics Office. Retrieved July 3, 2012, from http://www2.nature.nps.gov/stats/viewReport.cfm

National Park Service. (2012b). Climate change response program. Retrieved July 3, 2012, from http://nature.nps.gov/climatechange/

National Science Foundation [NSF]. (2006). Science and engineering indicators, 2006. Retrieved August 12, 2008, from http://128.150.4.107/statistics/seind06/pdfstart.htm

Nerlich, B., & Koteyko, N. (2009). Carbon reduction activism in the UK: Lexical creativity and lexical framing in the context of climate change. *Environmental Communication, 3*(2), 206–223.

Nisbet, M. C. (2009). Communicating climate change: Why frames matter to public engagement. *Environment, 51,* 514–518.

Nisbet, M. C., & Kotcher, J. E. (2009). A two-step flow of influence? Opinion-leader campaigns on climate change. *Science Communication, 30*(3), 328–354.

Noever, P., & Perrin, F. (Eds.) (2004). *Yves Klein: Air architecture.* Los Angeles, CA: MAK Center for Art and Architecture.

Nordhaus, T., & Shellenberger, M. (2007). *From the death of environmentalism to the politics of possibility.* Boston, MA: Houghton Mifflin.

Norgaard, K. M. (2011). *Living in denial: Climate change, emotions, and everyday life.* Cambridge, MA: MIT Press.

North, P. (2010) Eco-localisation as a progressive response to peak oil and climate change: A sympathetic critique. *Geoforum, 41*(4), 585–594.

Norton, T. (2007). The structuration of public participation: Organizing environmental control. *Environmental Communication, 1*(2), 146–170.

Norges offentlige utredninger [NOU]. (2006). Et klimavennlig Norge [A climate-friendly Norway]. *Norges offentlige utredninger, 18.* Retrieved August 14, 2010, from http://www.regjeringen.no/nb/dep/md/dok/nou-er/2006/nou-2006-18.html?id = 392348 (Available only in Norwegian).

Number of active users at Facebook over the years. (2012). Associated Press. Retrieved July 17, 2012, from http://finance.yahoo.com/news/number-active-users-facebook-over-years-214600186--finance.html

O'Connor, J. (1998). *Natural causes: Essays in ecological Marxism.* London, UK: Guilford Press.

O'Neill, S., & Hulme, M. (2009). An iconic approach to climate change. *Global Environmental Change, 19,* 402–410.

O'Neill, S., & Nicholson-Cole, S. (2009). "Fear won't do it": Promoting positive engagement with climate change through visual and iconic representations. *Science Communication, 30*(3), 355–379.

Ockwell, D., Whitmarsh, L., & O'Neill, S. (2009). Reorienting climate change communication for effective mitigation: Forcing people to be green or fostering grass-roots engagement? *Science Communication, 30*(3), 305–327.

Olausson, U. (2009). Global warming – Global responsibility? Media frames of collective action and scientific certainty. *Public Understanding of Science,* 18, 421–36.

OneClimate. (2010). Virtual Meeting. Retrieved September 10, 2010, from http://www.oneclimate.net/virtual-meeting

Örnebring, H., & Jönsson, A. M. (2004). Tabloid journalism and the public sphere: A historical perspective on tabloid journalism. *Journalism Studies, 5*(3), 283–295.

Owens, S. (2000). Engaging the public: Information and deliberation in environmental policy. *Environment and Planning A,* 32, 1141–1148.

Palmer, L. (2005). *Hays Woods/Oxygen bar project proposal materials.* Pittsburgh, PA: Carnegie Mellon University.

Paterson, M., & Stripple, J. (2007). Singing climate change into existence: On the territorialization of climate policymaking. In M. E. Pettenger (Ed.), *The social construction of climate change: Power, knowledge, norms and discourses* (pp. 149–172). Hampshire, UK: Ashgate.

Paterson, M., & Stripple, J. (2010). My Space: Governing individuals' carbon emissions. *Environment and Planning D: Society and Space, 28*(2), 341–362.

Pendleton, A. (2010). After Copenhagen. *Public Policy Research, 16*(4), 210–217.

People's Republic of China. (2004). Initial national communication on climate change. Beijing, China.

Peterson, M. N., Peterson, M. J., & Peterson, T. R. (2005). Conservation and the myth of consensus. *Conservation Biology, 19*(3), 762–767.

Peterson, M. N., Peterson, M. J., & Peterson, T. R. (2006). Why conservation needs dissent. *Conservation Biology, 20*(2), 576–578.

Peterson, T. R. (1997). *Sharing the Earth: The rhetoric of sustainable development.* Columbia: University of South Carolina Press.

Peterson, T. R., Peterson, M. N., Peterson, M. J., Allison, S. A., & Gore, D. (2006). To play the fool: Can environmental conservation and democracy survive social capital? *Communication and Critical/Cultural Studies, 3*(2), 116–140.

Pettenger, M. E. (2007). Introduction: Power, knowledge and the social construction of climate change. In R. Pettenger (Ed.), *The social construction of climate change: Power, knowledge, norms, discourses* (pp. 1–22). Hampshire, UK: Ashgate.

Phillips, L. (2011). *The promise of dialogue: The dialogic turn in the production and communication of knowledge*. Amsterdam, The Netherlands: John Benjamins.

Plotkin, S. (2004). Is bigger better? Moving toward a dispassionate view of SUVs. *Environment, 46*(9), 8–21.

Plows, A. (2008). Towards an analysis of the "success" of UK green protests. *British Politics*, 3, 92–109.

Polli, A. (2008). *Aer* online exhibition. Retrieved from http://www.greenmuseum.org/aer

Polli, A. (2010). Breathtaking: Media art and public participation in climate issues. *Proceedings of the Media Ecology Association*, Volume 11 (pp. 7–20). Retrieved from http://www.media-ecology.org/publications/MEA_proceedings/v11/1.%20Front%20matter.pdf

Portuguese Environment Agency. (2010). Fifth national communication to the United Nations Framework Convention on Climate Change. Second national communication in the context of the Kyoto Protocol – Revised Version. Amadora, Portugal: Portuguese Environment Agency.

Powell, M. C., & Colin, M. (2009). Participatory paradoxes: Facilitating citizen engagement in science and technology from the top-down? *Bulletin of Science, Technology & Society, 29*(4), 325–342.

Priest, S., H., Bonafadelli, H., & Rusanen, M. (2003). The "trust gap" hypothesis: Prediciting support for biotechnology across national cultures as a function trust in actors. *Risk Analysis, 23*(4), 751–766.

Radford, J. (Producer). (2001). *Rocky Mountain National Park: Spirit of the mountains.* [Motion picture]. United States: Rocky Mountain Nature Association.

Rancière, J. (2004). *The politics of aesthetics: The distribution of the sensible.* (G. Rockhill, Trans.). London, UK & New York, NY: Continuum.

Rancière, J. (2006). *Hatred of democracy.* London, UK: Verso.

Ravensbergen, D. (2009, December 11). In the heart of Copenhagen, a "people's climate forum." *The Tyee.* Retrieved March 15, 2010, from http://thetyee.ca/Blogs/TheHook/Environment/2009/12/12/PeoplesForum/

Rawls, J. (1971). *A theory of justice.* Cambridge, MA: Harvard University Press.

Rawls, J. (1993). *Political liberalism.* New York, NY: Columbia University Press.

Rayner, S. (2003). Democracy in the age of assessment: Reflections on the roles of expertise and democracy in public-sector decision making. *Science and Public Policy, 30*(3), 163–170.

Reggio, G. (Director & Producer). (1983). *Koyaanisqatsi: Life out of balance* [Motion picture]. United States: Island Alive, New Cinema.

Rein, M., & Schön, D. A. (1993). Reframing policy discourse. In F. Fischer & J. Forester (Eds.), *The argumentative turn in policy analysis and planning* (pp. 143–166). London, UK: Duke University Press.

Reisigl, M., & Wodak, R. (2009). The discourse-historical approach (DHA). In R. Wodak & M. Meyer (Eds.), *Methods of critical discourse analysis* (pp. 87–121). London, UK: Sage.

Renn, O. (2008). *Risk governance: Coping with uncertainty in a complex world.* London, UK: Earthscan.

Revkin, A. (2007). Climate change as news: Challenges in communicating environmental science. In J. DiMento & P. Doughman (Eds.), *Climate change: What it means for us, our children, and our grandchildren* (pp. 139–161). Cambridge, MA: MIT Press.

Rimbach, J. (2003). Oxygen bar in Atlantic City, NJ, feeds fresh air to tourists. *Free Republic.* Retrieved December 31, 2003, from http://www.freerepublic.com/focus/f-news/845765/posts

Rio Declaration on Environment and Development. (1992). Retrieved July 17, 2012, from http://www.unep.org/Documents.Multilingual/Default.asp?documentid = 78&articleid = 1163

Risbey, J. (2008). The new climate discourse: Alarmist or alarming? *Global Environmental Change, 18*(1), 26–37.

Rittsel, P. (2007) It riskerar att bli nästa miljöbov. *ComputerSweden*, p. 23. Retrieved November 17, 2007, from www.computersweden.idg.se

Rocky Mountain National Park. (2008, August). Never summer, ever summer [Educational program conducted at Rocky Mountain National Park].

Rogelj, J., Nabel, J., Chen, C., Hare, W., Markmann, K., Meinshausen, M., Schaeffer, M., Macey, K., & Hohne, N. (2010). Copenhagen accord pledges are paltry. *Nature, 464,* 1126–1128.

Rootes, C. (2004) Environmental movements. In D. A. Snow et al. (Eds.), *The Blackwell companion to social movements* (pp. 608–640). Oxford, UK: Blackwell.

Rootes, C. (2007). Acting locally: The character, contexts and significance of local environmental mobilizations. *Environmental Politics, 16*(5), 722–741.

Rose, N. S. (1999). *Powers of freedom: Reframing political thought.* Cambridge, UK: Cambridge University Press.

Rowe, G., & Frewer, L. J. (2004). Evaluating public-participation exercises: A research agenda. *Science, Technology, & Human Values, 29*(4), 512–556.

Rubin, Jeff. (2009). *Why your world is about to get a whole lot smaller: Oil and the end of globalization.* Toronto, Canada: Random House.

Rycroft, A. (2008). Greenpeace's Second Life. Is the media giant missing the virtual boat? Retrieved October 25, 2008, from http://sunshinecommunications.ca/html/modules.php?op = modload&name = News&file = article&sid = 109

Rycroft, A. (2011). Virtual public spheres: New political culture. Retrieved September 15, 2011, from http://sunshinecommunications.ca/html/modules.php?op = modload&name = News&file = article&sid = 112

Ryghaug, M. (2006). Some like it hot: Om klimaendring, forskning og journalistikk. *Norsk Medietidsskrift, 13*(3), 197–219.

Ryghaug, M. (2011). Obstacles to sustainable development: The destabilization of climate change knowledge. *Sustainable Development, 19*(3), 57–222.

Ryghaug, M., & Skjølsvold, T. M. (2010). The global warming of climate science: Climategate and the construction of scientific facts. *International Studies in the Philosophy of Science, 24*(3), 287–307.

Ryghaug, M., Sørensen, K. H., & Næss, R. (2011). Making sense of global warming: Norwegians appropriating knowledge of anthropogenic climate change. *Public Understanding of Science, 20*, 778–795.

Sales of canned oxygen to create fresh market for Seven-Eleven Japan. (2006). *Mainichi Daily News.* Retrieved May 14, 2006, from http://www.rapidnewswire.com/5146-cannedoxygen-0245.htm

Sandlin, J. A. and Walther, C. S. (2009). Moral identity formation and social movement learning in the voluntary simplicity movement. *Adult Education Quarterly, 59*(4), 298–317.

Santos, B., Sousa, Nunes, J. A., & Meneses, M. P. (2007). Introduction: Opening up the canon of knowledge and recognition of difference. In B. Sousa Santos (Ed.), *Another knowledge is possible: Beyond northern epistemologies* (pp. xvix–lxii). London, UK: Verso.

Saul, J. R. (2006). *The collapse of globalism and the reinvention of the world.* Toronto: Penguin Canada.

Saunders, C., & Price, S. (2009). One person's eu-topia, another's hell: Climate camp as a heterotopia. *Environmental Politics, 18*(1), 117–122.

Scandrett E. (2000). Community work, sustainable development, and environmental justice. *Scottish Journal of Community Work and Development* 6, 7–13.

Scandrett, E. (2007). Environmental justice in Scotland: Policy, pedagogy, and praxis. *Environmental Research Letters, 2*(4), 1–7.

Scandrett, E. (2010). Environmental justice in Scotland: Incorporation and conflict. In N. Davidson, P. McCafferty, & D. Miller (Eds.) *NeoLiberal Scotland: Class and society in a stateless nation* (pp. 183–201). Cambridge, UK: Cambridge Scholars Publishing.

Scandrett, E. (2012). Environmental justice: A question of social justice?. In G. Mooney, & G. Scott (Eds.), *Social justice and social policy in Scotland* (pp. 239–255). Bristol, UK: Policy Press.

Scandrett, E. Crowther, J., & Hemmi, A. (2009). Only the "tip of the iceberg": Making visible learning amongst cyberactivists in Friends of the Earth Scotland. *Proceedings of the ECREA conference 2008.* Barcelona, Spain: ECREA.

Scandrett, E., O'Leary, T., & Martinez, T. (2005). Learning environmental justice through dialogue. In *Proceedings of PASCAL conference: Making knowledge work* (pp. 345–352). Leicester, UK: National Institute of Adult Continuing Education.

Schlembach, R. (2011). How do radical climate movements negotiate their environmental and social agendas? A study of debates within the Camp for Climate Action (UK). *Critical Social Policy, 31*(2), 194–215.

Schneemann, C. (Director). (1964–1966). *Fuses* [Motion picture]. USA: Electronic Arts Intermix.

Scott, G., & Mooney, G. (2009). Poverty and social justice in the devolved Scotland: Neoliberalism meets social democracy? *Social Policy and Society, 3*(4), 379–389.

Scott, J. C. (1985). *Weapons of the weak: Everyday forms of peasant resistance.* New Haven, CT: Yale University Press.

Scottish Executive. (2002). *First minister's speech on environmental justice.* Retrieved January 30, 2007, from www.scotland.gov.uk/News/News-Extras/57

Scottish Government. (2004). *Do a little – Change a lot.* News release. Retrieved September 1, 2004, from www.scotland.gov.uk/News/Releases/2004/03/5132

Scottish Government. (2012). *£6.9 Million to Support Climate Reductions.* Retrieved June 25, 2012, from http://www.scotland.gov.uk/News/Releases/2012/03/climate-change13032012

Scottish Trades Union Congress [STUC]. (2010). A just transition to meet the challenge of climate change. Retrieved September 10, 2011, from http://www.stuc.org.uk/rebuilding-collective-prosperity.

Second Life climate conference. (2007). NewsCom.au. Retrieved December 14, 2007, from http://www.news.com.au/technology/climate-conference-goes-virtual/story-e6frfro0-1111115111651

Seel, B., Paterson, M., & Doherty, B. (2000). *Direct action in British environmentalism.* Oxford, UK: Routledge.

Segnit, N., & Ereaut, G. (2007). *Warm words II: How the climate story is evolving and the lessons we can learn from encouraging public action.* London, UK: Institute for Public Policy Research.

Shakhova, N., Semiletov, I., Salyuk, A., Yusupov, V., Kosmach, D., & Gustafsson, O. (2010). Extensive methane venting to the atmosphere from sediments of the East Siberian arctic shelf. *Science, 327,* 1246–1250.

Sharp, E. B., Daley, D. M., & Lynch, M. S. (2011). Understanding local adoption of climate change mitigation policy. *Urban Affairs Review, 47*(3), 433–457.

Silver, C. S. (1990). *One earth, one future: Our changing global environment.* Washington, DC: National Academies Press.

Silverman, D. (Director). (2007). *The Simpsons Movie* [Motion picture]. USA: 20th Century Fox.

Silverstone, R., & Hirsch, E. (Eds.) (1992). *Consuming technologies: Media and information in domestic spaces.* London, UK: Routledge.

Simms, A. (2005). *Ecological debt: The health of the planet and the wealth of nations.* London, UK: Pluto Press.

Simms, A. (2006, March). *The great emissions rights giveaway.* Feasta/ New Economics Foundation. Retrieved from www.feasta.org/documents/energy/emissions2006.pdf

Sklair, L. (2001). *The transnational capitalist class.* Oxford, UK: Blackwell.

Slocum, R. (2004). Polar bears and energy-efficient lightbulbs: Strategies to bring climate change home. *Environment and Planning D: Society and Space, 22*(3), 413–438.

Small, T. A. (2008). Equal access, unequal success: Major and minor Canadian parties on the Net. *Party Politics, 14*(1), 51–70.

Smith, C. (2009, July 23). Hello local, goodbye global: Relocalization movement gains momentum. *The Georgia Straight.* Retrieved March 15, 2010, from http://www.straight.com/article-241535/hello-local-goodbye-global

Smith, H. A. (2007). Disrupting the global discourse of climate change: The case of indigenous voices. In M. E. Pettenger (Ed.), *The social construction of climate change: Power, knowledge, norms, discourses* (pp. 197–217). Hampshire, UK: Ashgate.

Smith, W. D. (2007). Presence of mind as working climate change knowledge: A Totonac cosmopolitics. In M. E. Pettenger (Ed.), *The social construction of climate change: Power, knowledge, norms, discourses* (pp. 217–234). Hampshire, UK: Ashgate.

Sobel, D. (2004). *Place-based education: Connecting classrooms and communities.* Great Barrington, MA: Orion Society.

Solomon, D. (2007). Climate change's great divide. *The Wall Street Journal.* Retrieved September 12, 2007, from http://online.wsj.com/article/SB118955082446224332.html

Sørensen, K. H. (2006). Domestication: The enactment of technology. In T. Berker, M. Hartman, Y. Punie, & K. Ward (Eds.), *Domestication of media and technology* (pp. 40–61). Maidenhead, UK: Open University Press.

Sørensen, K. H., Aune, M., & Hatling, M. (2000). Against linearity: On the cultural appropriation of science and technology. In M. Dierkes, & C. von Grote (Eds.), *Between understanding and trust: The public, science and technology* (pp. 237–257). Amsterdam, The Netherlands: Harwood.

Speth, J. G. (2004). *Red sky at morning: America and the crisis of the global environment; A citizen's agenda for action.* New Haven, CT: Yale University Press.

Stavins, R. N. (2008). Addressing climate change with a comprehensive US cap-and-trade system. *Oxford Review of Economic Policy, 24*(2), 298–321.

Stedman, R. C. (2003). Sense of place and forest science: Toward a program of quantitative research. *Forest Science, 49*(6), 822–829.

Steel, D. (2000). Polar bonds: Environmental relationships in polar regions. *Environment and Behavior, 32,* 796–816.

Steiner, J., Bächtiger, A., Spörndli, M., & Steenbergen, M. R. (2005). *Deliberative democracy in action: Analyzing parliamentary discourse.* Cambridge, UK: Cambridge University Press.

Step It Up. (2007). Step it up 2007: National days of climate action. Retrieved November 8, 2010, from http://stepitup2007.org/

Sterman, J. D., & Sweeney, L. B. (2007). Understanding public complacency about climate change: Adults' mental models and climate change violate conservation of matter. *Climatic Change, 80,* 213–238.

Stewart, D. W., Shamdasani, P. N., & Rook, D. W. (2007). *Focus groups: Theory and practice.* Thousand Oaks, CA: Sage.

Stirling, A. (2006). Precaution, foresight, and sustainability: Reflection and reflexivity in the governance of science and technology. In J.-P. Voß, D. Bauknecht, & R. Kemp (Eds.), *Reflexive Governance for Sustainable Development* (pp. 225–272). Cheltenham/Northampton, UK: Edward Elgar.

Stirling, A. (2008). "Opening up" and "Closing down": Power, participation, and pluralism in the social appraisal of technology. *Science, Technology, & Human Values, 33*(2), 262–294.

Stoll-Kleemann, S., O'Riordan, T., & Jaeger, C. C. (2001). The psychology of denial concerning climate mitigation measures: Evidence from Swiss focus groups. *Global Environmental Change, 11,* 107–117.

Straight editorial staff. (2009, May 7). The Straight slate for progressive change, *The Georgia Straight*. Retrieved March 15, 2010, from http://www.straight.com/article-219083/straight-slate-progressive-change

Stratospheric ozone depletion 10 years after Montreal. (1997). *Natural Science*. Retrieved November 1, 2007, from http://naturalscience.com/ns/cover/cover4.html

Strauss, A., & Corbin, J. (1998). *Basics of qualitative research: Techniques and procedures for developing grounded theory*. Thousand Oaks, CA: Sage.

Suzuki, D., & Moola, F. (2009a, January 6). David Suzuki: Canadians must take personal responsibility for climate change. *The Georgia Straight*. Retrieved March 15, 2010, from http://www.straight.com/article-178473/david-suzuki-canadians-must-take-personal-responsibility-climate-change

Suzuki, D., & Moola, F. (2009b, October 27). David Suzuki: Copenhagen climate is crucial. *The Georgia Straight*. Retrieved March 15, 2010, from http://www.straight.com/article-266324/david-suzuki-copenhagen-climate-summit-crucial

Suzuki, D., & Moola, F. (2009c, November 3). David Suzuki: Inaction on climate change comes with a huge price tag. *The Georgia Straight*. Retrieved March 15, 2010, from http://www.straight.com/article-268470/david-suzuki-inaction-climate-change-comes-huge-price-tag

Suzuki, D., & Moola, F. (2009d, May 5). David Suzuki: Obama sets example for endangered species protection. *The Georgia Straight*. Retrieved March 15, 2010, from http://www.straight.com/article-219001/david-suzuki-obama-sets-example-endangered-species-protection

Swyngedouw, E. (1997). Neither global nor local: "Glocalisation" and the politics of scale. In K. Cox (Ed.), *Spaces of globalisation: Reasserting the power of the local* (pp. 137–166). London, UK: Guildford Press.

Swyngedouw, E. (2010). Apocalypse forever? Post-political populism and the spectre of climate change. *Theory, Culture & Society, 27*(2–3), 213–232.

Taylor, M., Kent, M. L., & White, W. J. (2001). How activist organizations are using the Internet to build relationships. *Public Relations Review*, 27, 263–284.

The Compensators*. (2011). Tag archive for cancellations. Retrieved December 10, 2011, from Please replace highlighted text with http://thecompensators.org/tag/cancellations/

Thomas, A. J. (2002, February). *Inschrift des Krieges*. Retrieved August 6, 2010, from Senses of Cinema: http://archive.sensesofcinema.com/contents/cteq/01/19/bilder.html

Thomashow, M. (2002). *Bringing the biosphere home: Learning to perceive global environmental change*. Cambridge, MA: MIT Press.

Thornes, J. E. (2008). Cultural climatology and the representation of the sky, atmosphere, weather, and climate in selected art works of Constable, Monet, and Eliasson. *Geoforum*, 38, 570–580.

Thornes, J. E., & Randalls, S. (2007). Commodifying the atmosphere: Pennies from heaven? *Geografiska Annaler*, 89(4), 273–285.

TNS Gallup. (2010). Tror fremdeles på menneskeskapte klimaendringer. Retrieved August 14, 2010, from http://www.tns-gallup.no/?did = 90 92915

Toker, C. W. (2005). The deliberative ideal and co-optation in the Georgia Port Authority's stakeholder evaluation group. *Environmental Communication Yearbook*, 2, 19–48.

Trenz, H.-J. (2009). Digital media and the return of the representative public sphere, *Javnost: The Public*, 16(1), 33–46.

Trumbo, C. W. (1995). Longitudinal modeling of public issues. An application of the agenda-setting process to the issue of global warming. *Journalism and Mass Communication Monographs* 152, 1–57.

Tsoukas, H. (1999). David and Goliath in the risk society: Making sense of the conflict between Shell and Greenpeace in the North Sea. *Organization*, 6(3), 499–528.

Turkle, S. (1995). *Life on the screen: Identity in the age of the Internet*. New York, NY: Simon & Schuster.

Turner, C. (2007). *The geography of hope: A tour of the world we need.* Toronto, Canada: Random House.

The Tyee. (2010). Magazine Association of British Columbia. Retrieved October 10, 2010, from http://magsbc.com/magazines/tyee

U.S. Central Intelligence Agency. (2011). *The world factbook.* Retrieved March 10, 2011, from https://www.cia.gov/library/publications/the-world-factbook/

U.S. Central Intelligence Agency. (2012). *The world factbook.* Retrieved July 10, 2012, from https://www.cia.gov/library/publications/the-world-factbook/

Ulysses. (2012). The ULYSSES Web tutorial on participatory integrated assessment. Retrieved July 27, 2012, from www.jvds.nl/ulysses

United Nations Environment Programme [UNEP] & Futerra. (2005). *Communicating sustainability: How to produce effective public campaigns.* Paris: UNEP.

United Nations Framework Convention on Climate Change [UNFCCC]. (2007). *Kyoto Protocol.* Retrieved November 10, 2007, from http://unfccc.int/kyoto_protocol/items/2830.php

United Nations Framework Convention on Climate Change [UNFCCC]. (1992). First steps to a safer future: Introducing the United Nations Framework Convention on Climate Change. Retrieved July 16, 2012, from http://unfccc.int/essential_background/convention/items/6036.php

United Nations Framework Convention on Climate Change [UNFCCC]. (1999). Review of the implementation of commitments and of other provisions of the convention: UNFCC guidelines on reporting and review. Retrieved July 16, 2012, from unfccc.int/resource/docs/cop5/07.pdf

United Nations Framework Convention on Climate Change [UNFCCC]. (2003). *Report of the Conference of the Parties on its eighth session, held at New Delhi from 23 October to 1 November 2002: Addendum. Part two: Action taken by the Conference of the Parties at its eighth session.* Retrieved July 16, 2012 from unfccc.int/resource/docs/cop8/07a01.pdf

United Nations Framework Convention on Climate Change [UNFCCC]. (2008). *Report of the Conference of the Parties on its thirteenth session, held in Bali from 3 to 15 December 2007: Addendum. Part two: Action taken by the Conference of the Parties at its thirteenth session.* Retrieved July 16, 2012 from unfccc.int/resource/docs/2007/cop13/eng/06a01.pdf

United Nations Framework Convention on Climate Change [UNFCCC]. (2012a). *National Communications Annex I.* Retrieved July 18, 2012, from http://unfccc.int/national_reports/annex_i_natcom_/items/1095.php

United Nations Framework Convention on Climate Change [UNFCCC]. (2012b). *National Communications and Biennial Update Reports from Non–Annex I Parties.* Retrieved July 18, 2012, from http://unfccc.int/national_reports/non-annex_i_natcom/items/2716.php

United Nations Framework Convention on Climate Change [UNFCCC]. (2012c). *National reports.* Retrieved June 20, 2012 from http://unfccc.int/national_reports/items/1408.php

United Nations Framework Convention on Climate Change [UNFCCC]. (n.d.). Decision -/CP.10. [Advance unedited version]. Status of, and ways to enhance, implementation of the New Delhi work programme on Article 6 of the Convention. Retrieved July 16, 2012, from http://unfccc.int/cooperation_and_support/education_and_outreach/items/2529.php

United Nations Statistics Division. (2012). Millennium development goals indicators. Retrieved July 20, 2012, from http://mdgs.un.org/unsd/mdg/Data.aspx and http://mdgs.un.org/unsd/mdg/SeriesDetail.aspx?srid = 749&crid =

United States Congress. (1997, July). *The joint economic committee study.* Retrieved November 1, 2007, from http://www.house.gov/jec/cost-gov/regs/cost/emission.html

United States Department of State. (2010). *US climate action report: Fifth national communication of the United States of America under the United Nations Framework Convention on Climate Change.* Washington, DC: Global Publishing Services.

United States Geological Survey [USGS]. (2007). Climate change in mountain ecosystems. Retrieved August 12, 2008, from http://www.nrmsc.usgs.gov/research/global.htm

USCAN Climate Action Network. (2010). *Who's on board with the Copenhagen Accord?* Retrieved March 1, 2011, from http://www.usclimatenetwork.org/policy/copenhagen-accord-commitments

Van Beukering, P., & Vellinga, P. (1996). Climate change: From science to global politics. In P. Sloep & A. Blowers (Eds.), *Environmental policy in an international context: 2. Conflicts* (pp. 187–215). London, UK: Arnold.

Van de Kerkhof, M. (2006). Making a difference: On the constraints of consensus building and the relevance of deliberation in stakeholder dialogues. *Policy Sciences, 39,* 279–299.

van Dijk, T. (1988). *News as discourse.* Hillsdale, NJ: Laurence Erlbaum.

Vermeer, M., & Rahmstorf, S. (2009, December 7). Global sea level linked to global temperature. *Proceedings of the National Academy of Sciences of the United States, 106* (51), 21527–21532.

Vitterso, J., Vorkinn, M., & Vistad, O. I. (2001). Congruence between recreational mode and actual behavior: A prerequisite for optimal experiences? *Journal of Leisure Research, 33*(2), 137–159.

Wahlström, B. (2007). *Guide till det virtuella samhället.* Stockholm, Sweden: SNS-förlag.

Wallström, M. (2007). IBMs programguru har licens att döda. *ComputerSweden,* p. 7. Retrieved November 21, 2007, from www.computersweden.idg.se

Wang, T., & Watson, J. (2008). China's carbon emissions and international trade: Implications for post-2012 policy. *Climate Policy, 8*(6), 577–587.

Webler, T., Tuler, S., & Krueger, R. (2001). What is a good public participation process? Five perspectives from the public. *Environmental Management, 27,* 435–450.

Weingart, P. (1998). Science and the media. *Research Policy, 27,* 869–879.

Weingart, P., Engels, A., & Pansegrau, P. (2000). Risks of communication: Discourses on climate change in science, politics, and the mass media. *Public Understanding of Science*, 9, 261–283.

Westling, M. (2007). Expanding the public sphere: The impact of Facebook on political communication. *The New Vernacular*. Retrieved August 15, 2010, from www.thenewvernacular.com

Whiteside, K. H. (2006). *Precautionary politics: Principle and practice in confronting environmental risk*. Cambridge, MA: MIT Press.

Whitmarsh, L., O'Neill, S., & Lorenzoni, I. (Eds.). (2010). *Engaging the public with climate change: Behaviour change and communication*. London, UK: Earthscan.

Wickham, T. D. (2001). Attachments to place and activities: The relationships of psychological constructs to customer satisfaction. *Dissertation Abstracts*, 16, 3348.

Wiener, D. (2000, August). Bird's-Eye View. *American Cinematographer*, pp. 92–116.

Wilkinson, M., & Scandrett, E. (2003). A popular education approach to tackling environmental injustice and widening participation. *Concept*, 13(1/2), 11–16.

Williams, C. B., & Gulati, G. J. (2008). The political impact of Facebook: Evidence from the 2006 midterm elections and 2008 nomination contest. *Politics & Technology Review*, 1, 11–21.

Williams, D. R., & Vaske, J. J. (2003). The measurement of place attachment: Validity and generalizability of a psychometric approach. *Forest Science*, 49(6), 830–839.

Williams, R. (1989). *Resources of hope*. London, UK: Verso.

Willoquet-Maricondi, P. (2010). Shifting paradigms: From environmentalist films to ecocinema. In P. Willoquet-Maricondi (Ed.), *Framing the World: Explorations in Ecocriticism and Film* (pp. 43–61). Charlottesville: University of Virginia Press.

Wilson, J. (2009, December 4). Jessica Wilson: Stephen Harper goes to Copenhagen as climate enemy number one. *The Georgia Straight*. Retrieved March 15, 2010, from http://www.straight.com/article-273798/vancouver/jessica-wilson-stephen-harper-goes-copenhagen-climate-enemy-number-one

Wilson, K. (1995). Mass media as sources of global warming knowledge. *Mass Communication Review, 22*(1–2), 75–89.

Wodak, R., & M. Meyer (Eds.). (2009). *Methods of critical discourse analysis.* London, UK: Sage.

Wodak, R. (2011). *The discourse of politics in action: Politics as usual.* Basingstoke, UK: Palgrave.

Wojcik, D. (1996). Embracing doomsday: Faith, fatalism, and apocalyptic beliefs in the nuclear age. *Western Folklore,* 55, 297–330.

Wood, P. (2002). *Movements in modern art: Conceptual art.* Delano, NY: Greenridge Editions.

World Wide Views [WWViews]. (n.d.). World Wide Views on Global Warming: WWViews project description. Retrieved July 19, 2011, from http://www.wwviews.org/node/6

Wyler, R. (2009, February 26). Peak oil means sooner or later we'll wake up to a new normal. *The Georgia Straight.* Retrieved March 15, 2010, from http://www.straight.com/article-203816/rex-weyler-peak-oil-means-sooner-or-later-well-wake-new-normal

Wynne, B. (1995). Public understanding of science. In S. Jasanoff, G. E. Markle, J. C. Petersen, & T. Pinch (Eds.), *Handbook of science and technology studies* (pp. 361–88). Thousand Oaks, CA: Sage.

Yale Project on Climate Change & the George Mason University Center for Climate Change Communication. (2009). Climate change in the American mind: American's climate change beliefs, attitudes, policy preferences, and actions. Retrieved March 29, 2009, from http://www.climatechangecommunication.org/images/files/Climate_Change_in_the_American_Mind.pdf

Yang, G., & Calhoun, C. (2007). Media, civil society, and the rise of a green public sphere in China. *China Information, 21*(2), 211–236.

Zavestoski, S., Shulman, S., & Schlosberg, D. (2006). Democracy and the environment on the Internet electronic citizen participation in regulatory rulemaking. *Science, Technology, & Human Values, 31*(4), 383–408.

Zehr, S. (2009). An environmentalist/economic hybrid frame in US press coverage of climate change, 2000–2008. In T. Boyce & J. Lewis (Eds.), *Climate Change and the Media* (pp. 80–91). New York, NY: Peter Lang.

Zero Carbon Britain Project. (2010). *ZeroCarbonBritain2030: A new energy strategy*. Machynlleth, UK: Centre for Alternative Technology. Retrieved September 10, 2011, from http://www.zerocarbonbritain.org/resources/cats-previous-zcb-reports.

Zia, A., & Todd, A. M. (2010). Evaluating the effects of ideology on public understanding of climate change science: How to improve communication across ideological divides? *Public Understanding of Science, 19*(6), 743–776.

Ziser, M., & Sze, J. (2007). Climate change, environmental aesthetics, and global environmental justice studies. *Discourse, 29*(2–3), 384–410.

Žižek, S. (1999). *The ticklish subject: The absent centre of political ontology*. London, UK: Verso.

Index

11th Hour, The, 21, 99, 115, 117, 119
350.org, 23, 196, 201–202, 205–211, 214–217

Aer, 234–236, 238, 242–243
Air Architecture, 231–232
Ann Arbor, 208, 212, 214
Annex I countries, 134–135, 138, 148
Antananarivo, 204–205, 212, 214
activism, 6, 11, 21–23, 26, 82, 92, 110–111, 119, 165, 169, 177, 183, 235, 239, 241, 243–245, 248–249, 253, 255, 261, 264–270, 272, 311, 314–316
aerial shot, 88–90, 93, 113, 116
agonistic pluralism, 8, 14–16, 18, 26–27, 219, 309, 311–313, 318
alternative media, 16, 25–26, 166, 179, 247, 250–253, 255, 257–259, 263, 265–266, 268–269, 271–274, 311–312, 314, 317
apocalypse, 4, 237, 257–258, 271, 273–274, 285
argument, 4, 12–13, 17, 38, 44, 90, 92, 100, 103, 108–109, 115–117, 125, 149, 190, 202, 265, 286, 303
art, artistic, 24, 88, 91–92, 94, 107, 221–222, 230–235, 238, 240–245, 311, 317
attitude, 34, 141, 228

awareness, 20–21, 24, 35, 54, 61, 81–82, 109, 112, 127, 129–131, 133–134, 137, 140–141, 143, 146, 149–151, 153–157, 159, 204, 208, 214, 234, 242, 256, 259, 274, 316

behavior, 8–9, 33–34, 49, 51, 53, 63–64, 66, 68, 83–84, 123, 141, 159, 175–176, 270–271, 283, 308
Bonaqua Condensation Cube, 233, 244
bottom-up, 23, 27, 35, 217, 280
British Columbia, 259–261, 266, 273

Camps for Climate Change
Canada, 184–185, 251, 255–256, 259, 267, 299, 315
cap-and-trade, 222–223, 226–227
capitalist accumulation, 281
Caracas, 208, 212
carbon credits, 226–227, 287
carbon emissions, 108, 145, 183, 221, 224
carbon trading, 223, 227–229, 280, 288, 311
carbon web, 235
Catia la Mar, 210, 214
China, 18, 22, 106, 134, 148–152, 193, 224
chlorofluorocarbons, 223–224, 243

citizen, 2–3, 6–8, 10–12, 16–17, 22, 24, 26–27, 38, 52, 56, 59, 66, 72, 84, 115, 123–124, 143, 152, 156–157, 164–165, 188–189, 194–195, 197–198, 214, 229, 264, 266, 307–308, 314
civic engagement, 11, 184, 248–250, 271–272, 274
class, 26, 39, 206, 212, 277, 279–282, 285–287, 289–296, 298–304, 312
Climate Challenge Fund, 282, 295
climate change denial, 284
climate justice, 6, 26, 53, 206, 212, 218, 235, 257, 260, 267–268, 272, 280, 282–286, 289, 291–295, 312–313, 317
climate treaty, 201, 270
commodification, 24, 230–231, 239, 280
communication, 2–3, 6–9, 11–12, 15–20, 22–24, 27, 31, 60–68, 70, 74, 77, 81, 85–86, 89–90, 103–104, 112, 114, 117–118, 128, 133, 135–136, 142, 150, 164, 166–171, 174, 176–177, 179, 181–190, 195, 197, 204–205, 238, 247–248, 250, 258, 272, 283, 292, 299, 301, 307–309, 312–315, 317–318
consensus, 4, 10–16, 39, 41–42, 128, 142, 248, 256, 314, 316–317
conflict, 4, 13–14, 127–128, 139, 277, 290, 299, 301, 303
Copenhagen, 6, 195, 197, 199–200, 217, 249, 251, 254, 259–260, 265–269, 273, 310, 315
corporate ownership, 224

deliberation, 10–12, 23, 40, 52, 125, 164, 166–168, 170, 183, 186–190, 310–311, 317

democracy, 2–3, 12–15, 23, 25, 27, 71, 125–127, 146, 151, 159, 164, 167–168, 188–190, 248–250, 253–254, 262, 264, 271, 282, 285–286, 291, 293, 302, 310–313, 317–318
demonstration, 83, 105, 197, 217
Denver, 208, 214
decision, 84, 142, 164, 205, 213
depoliticization, 5, 313
dialogical, 11, 27, 148, 282, 292, 296, 310
direct action, 27, 278, 285, 291, 298, 300
discourse, 4–5, 7, 12–13, 16–18, 20, 59–60, 62, 64, 66–68, 71, 73–75, 78, 82, 84–86, 100–101, 105–106, 119, 124–126, 135–137, 148, 163–165, 167–171, 174, 176, 179, 181, 183, 186–188, 230, 248, 251–252, 259, 265, 272–274, 279–280, 282, 284–290, 292–297, 303, 318
discourse analysis, 135, 251–252
disenfranchised, 206, 290
dissensus, 25, 311
documentary, 20–21, 87–88, 91, 97, 101–102, 105–106, 110–111, 117–120, 140
domestication, 33, 35–39, 41, 43, 46, 48–49, 51–52, 54–55

epistemic community, 278–279
European Union, 15, 23, 145, 194, 212, 221, 229, 240, 242
Earth in the Balance, 228
ecological economics, 225
ecological modernisation, 289

education, 3, 9, 19–20, 27, 39–40, 65, 67–68, 73, 75–76, 80, 105, 129–131, 133–134, 137, 139–143, 146, 149–154, 156–157, 159–160, 205, 207, 235, 277, 282, 291–293, 295–296, 302, 304, 312
environmental activism, 82, 119
environmental justice, 26, 206, 212, 218, 283, 286, 289–290, 292–294
everyday life, 31, 33, 35–38, 41–44, 48–49, 51–56, 89, 164
Everything's Cool, 21, 99, 110–111, 115, 117, 119

Facebook, 23, 167–174, 176–179, 186, 188–189, 196, 310
Federal Communications Commission, 224–225, 229
focus groups, 40
fossil fuel, 31, 38, 107, 229, 293, 299, 303, 316
frame, 9, 25, 43, 51, 102–105, 111, 129, 131, 165, 170, 172, 177, 188, 202, 249, 252, 256–257, 259, 268, 285, 311
Friends of the Earth, 9, 278, 281–282, 287, 292

generative analysis, 60, 62
Giddens, 24, 31–32, 45, 54, 126, 194, 197–199, 212–213, 217, 279
Glacier National Park, 63, 78–83
Gore, Al, 101–103, 111, 140, 154, 228
governance, 10, 12, 18, 23, 37, 48, 126–127, 131, 141, 157, 163–166, 170, 177, 187–188, 190, 194, 281, 310

government, 8, 13, 24, 41–42, 47, 49, 60, 89, 109, 123–124, 141–145, 147, 149–152, 154, 156, 158, 197, 199–202, 205, 212–213, 215, 226–228, 242–243, 249, 255, 257–261, 266, 269, 283–284, 286, 293–295, 309
governmentality, 5, 9, 160
grassroots, 8, 101, 110, 215, 217, 244, 268–269, 291, 311
Great Emissions Rights Giveaway, The, 229
greenhouse gas, 1, 6, 18, 22, 25, 40, 48, 60, 77, 79, 124, 134, 138, 141–142, 145–146, 148–149, 151, 156, 194, 204, 216–217, 221–222, 225–226, 229, 249, 251, 278, 315, 318
Greenpeace, 9, 170, 177–179, 181, 183–186, 264, 278

Habermas, Jürgen, 165
hegemony, 3–5, 7, 10, 13, 16, 24–25, 27, 233, 245, 285–286, 303, 311–312, 318
human environment, 88, 278

Inconvenient Truth, An, 21, 99, 101, 103, 107, 109–111, 115, 117, 140
India, 18, 22, 134, 149, 151–153, 155
Inert Gas Series, 232
information deficit
Internet, 7, 22–23, 150, 167–170, 179–180, 189, 195–197, 199, 217, 242–243, 310–311, 314–315, 317
Intergovernmental Panel on Climate Change (IPCC), 41, 83, 140, 242, 274, 278, 280, 312

knowledge, 10, 22, 25–27, 32–38, 40, 42–47, 49, 52–56, 83, 106, 112, 118, 127, 136, 139, 141–143, 150–155, 159, 172, 174, 176, 196, 200, 213, 215, 247, 278, 280–282, 288, 291–292, 296–298, 301, 312
Kyoto Protocol, 18, 130–131, 134, 138, 145, 147, 157, 221–222, 227, 242

landscape, 3, 18, 20–21, 59–60, 62–68, 70–76, 78–85, 88–90, 92, 168, 241, 309
lifestyle, 8, 20, 25, 42, 60, 175–176, 251–253, 257, 264, 283, 285, 294, 308
long shot, 88, 93

Madagascar, 197, 199, 204–206, 212–214
Maracaibo, 210, 214
media, 3, 5, 7, 12, 16, 19–20, 25–26, 33, 36–37, 43–46, 53, 55–56, 86, 92, 112, 114–115, 136, 140, 150, 153, 155–156, 163, 165–171, 174, 176–180, 186–190, 196, 198–199, 210, 218, 225, 234, 236, 239, 245, 247–253, 255–259, 263–266, 268–269, 271–274, 285, 308–312, 314, 317
McKibben, Bill, 110, 113, 115, 195, 316
Montreal Protocol, 223, 227
Mouffe, Chantal, 4, 6, 12–13, 233

narrative, 44, 55, 271, 283, 285, 287
national parks, 19–20, 59–60, 63–69, 78, 83, 86
National Park Service, 59, 63, 308
neoliberalism, 282, 286–287

nongovernmental organization (NGO), 1, 8, 23, 27, 125, 131, 147–148, 154–155, 166, 169–170, 177, 179, 186, 196–197, 199, 210, 217, 268, 279, 293, 310, 316
non–Annex I countries, 134–135, 148
norm, 92
Norway, 18, 31–33, 41–42, 45, 47, 50

Obama, Barack, 184, 255–256, 315–316
oxygen bar, 230–231, 233–234, 244

place attachment theory
place-based communication, 66
policy, 4, 8–12, 16, 20–24, 32–37, 41–43, 47, 50, 53, 56, 60, 63, 66, 83, 86, 89, 91, 93–94, 111–112, 117–118, 124–126, 128–129, 132, 134, 137, 139, 142, 144, 146, 148, 152, 157–158, 160, 164–165, 169, 193–198, 200, 204, 212, 214, 216–217, 229, 235, 250, 253, 256–260, 269, 272, 278–284, 286–288, 291, 293–298, 303, 307–313, 315–318
political, the, 1–4, 6, 8–10, 16, 23–26, 32, 37, 47, 60, 66, 68, 86, 101, 110, 164, 167, 187–188, 193, 211, 215, 222, 226, 228, 248–249, 252, 254, 259, 264, 267, 311–314
political contestation, 221
political mobilization, 9, 248, 311

politics, 1–4, 6–8, 12–19, 22–27, 31–33, 38, 44, 46–47, 53–54, 86–88, 92, 95, 101, 110, 114–115, 123–125, 129, 132, 134–135, 137, 148–149, 155, 164, 168, 187–188, 190, 193, 204, 212, 217–219, 235, 242, 245, 247–253, 255–259, 261, 264–266, 268–272, 274, 300, 307–315, 317, 319
pollutant, 24, 34–35, 99, 178, 222, 225–227, 235–236, 239, 243, 286, 293–294, 299
popular education, 3, 27, 277, 282, 291–293, 302, 304
Portugal, 22, 134, 145–146, 148, 158
postpolitical, 3–4, 15, 24, 279
poverty, 151, 153, 277, 281–282, 290, 294–298, 302
progressive, 87, 183, 253, 255, 260, 315
property rights, 225–226, 229, 280, 287–288
protest, 7, 24, 204, 267–268, 277, 299, 310, 316
public awareness, 20, 35, 112, 129–131, 133–134, 137, 140–141, 143, 146, 149–151, 155–157, 256, 274, 316
public consultation, 10, 147–148, 158
public engagement, 7–8, 10, 18, 20–22, 25, 27, 29, 59–60, 66, 86, 121, 125, 127, 136, 143, 158–159, 196, 219, 248, 258, 309
public expertise, 35, 119, 132, 196, 316
public participation, 8, 10–12, 18, 22, 25, 27, 86, 121, 123–137, 140, 145–146, 151, 154–155, 157–158, 164, 187, 193–194, 239, 309–311, 313

public sphere, 4, 7, 10, 62, 164–169, 171, 176, 179, 183–184, 186–190, 272, 314, 317

rhetoric, 21, 87, 102, 105, 176, 241, 258
Rocky Mountain National Park, 63, 68, 70–74, 78

Salzberg, 202, 214
San Juan, 207, 212
Sana'a, 210
Scotland, 282, 286, 288, 292–293, 295, 299–301, 303
Seattle, 169, 207, 213
Second Life, 23, 163, 168, 170, 179–189, 310–311
sense-making, 32–33, 36–38, 43–46, 49, 53–55
social democracy, 282, 285–286, 293
social justice, 53, 260, 264, 272, 282, 284–286, 288, 290, 293, 312
social marketing, 8–9, 12, 18, 20, 27, 29, 313
social movement, 6, 278, 291, 298, 315
social networks, 39, 168
social system, 61, 198
socialist transformation, 282, 289
state commitment, 123
Step It Up, 195, 215–216
Stephansplatz, 200, 202–203
structural changes, 253, 313
structuration, 24, 197–199, 217–218
subaltern interests, 279–282, 289, 292, 304
subsidy, 227

sustainable, 4–6, 10, 32, 38, 42, 67, 73, 77, 80, 93, 106, 109, 120, 133–134, 143, 146, 149, 151–153, 156, 158, 165, 181–183, 186, 188–190, 206, 229, 253, 258–259, 261–264, 272, 279, 281, 284, 287–288, 297–298, 307, 313, 317
Sweden, 18, 179

technomanagerial, 7, 10, 24, 26, 311, 313
top-down, 9, 12, 17, 19, 27, 128, 217, 229, 244, 247
Tragedy of the Commons, 225
Tumero, 209
Tuvalu, 22, 134, 155–156

United Kingdom, 22, 134, 142, 259
United Nations, 6, 9, 32, 124, 137–138, 182–183, 193, 195, 207, 215, 217, 222–223, 227, 242, 278–280, 309

United Nations Framework Convention on Climate Change (UNFCCC), 6, 18, 21–22, 123–125, 129–135, 137–138, 141, 146, 155–158, 222, 309, 315
USA, 109

Venezuela, 197, 199, 208–210, 212–214

Wales, 144–145, 303
Wall Street Journal, The, 227
Weather Makers, The, 227
Western Climate Initiative, 260
working class, 280, 286, 291, 294, 299, 303

Yellowstone National Park, 63, 75–76
Yemen, 18, 106, 197, 199, 210–211, 213–215

About the Contributors

Anabela Carvalho (PhD, University College London) is an associate professor in the Department of Communication Sciences at the University of Minho, Portugal. Her research focuses on various forms of environment, science, and political communication with a particular emphasis on climate change. She is editor of *Communicating Climate Change: Discourses, Mediations and Perceptions* (2008), *As Alterações Climáticas, os Media e os Cidadãos* (2011), *Citizen Voices: Performing Public Participation in Science and Environment Communication* (with L. Phillips & J. Doyle, 2012) and of two journal special issues. Anabela Carvalho is associate editor of *Environmental Communication: A Journal of Nature and Culture* and is on the board of directors of the International Environmental Communication Association. She is also a cofounder, former chair, and currently vice chair of the Science and Environment Communication Section of ECREA.

Jim Crowther is a senior lecturer at the University of Edinburgh. He has undertaken research on the role of digital technologies in learning and action for environmental justice in Scotland. He is coordinator of the international Popular Education Network (PEN), which is a global network of academics and researchers with an interest in promoting popular education research and pedagogy. His most recent book with Budd Hall, Darlene Clover, and Eurig Scandrett, *Learning and Education for a Better World: The Role of Social Movements*, is published by Sense. He is currently the editor of the international journal *Studies in the Education of Adults*.

Andrea Feldpausch-Parker is an assistant professor in the Department of Environmental Studies at the State University of New York College of Environmental Science and Forestry. Her research focuses on environmental communication, with interests in conservation of natural resources through communication and cooperative learning among scientists, policy makers, and the general public. Her current focus is on climate change mitigation and adaptation strategies; additional interests include natural-resources management and policy, energy policy, public participation in environmental decision making, and social movements.

Joyeeta Gupta is a professor of climate change law and policy at the Vrije Universiteit Amsterdam and of water law and policy at the UNESCO-IHE Institute for Water Education in Delft. Her expertise is in the area of environmental law and politics. She is editor in chief of *International Environmental Agreements: Politics, Law, and Economics* and is on the editorial board of several journals. She was lead author in the Intergovernmental Panel on Climate Change and the Millennium Ecosystem Assessment. She is on the scientific steering committees of many international programs, including the Global Water Systems Project. She has authored and edited several books on climate change; among them are *The Climate Change Convention and Developing Countries—From Conflict to Consensus?*, *Our Simmering Planet: What to do About Global Warming*, *Climate Change and the Kyoto Protocol: The Role of Institutions and Instruments to Control Global Change*, *Issues in International Climate Policy: Theory and Policy*, and *Climate Change and European Leadership: A Sustainable Role for Europe*.

Shane Gunster is an associate professor in the School of Communication at Simon Fraser University, Canada. His current research focuses on media coverage of climate change and energy, as well as activist and advocacy communication about environmental politics. He is the author of *Capitalizing on Culture: Critical Theory for Cultural Studies* and his work has recently been published in the *Canadian Journal of Communication*, the *Canadian Journal of Political Science*, *Television and New Media*, and *Ethics and the Environment*.

Helen Hughes is a senior lecturer in film studies at the University of Surrey. She has published articles and chapters on West German cinema, Austrian experimental film (Valie Export, Ferry Radax), Kafka adaptations, the German Democratic Republic and new Austrian documentary cinema, and ecodocs (Niklaus Geyrhalter, Hubert Sauper, Rob Stewart). She coedited *Deutschland im Spiegel seiner Filme* (2000) and translated Alexander Kluge's *Cinema Stories* (2007) with Martin Brady. She is completing a monograph on environmental documentary film in the 21st century.

Anna Maria Jönsson is an associate professor in the Department of Culture and Communication at Södertörn University, Sweden. Her doctoral thesis from 2004, *Samma nyheter eller likadana?* ("Same news or similar?"), analyzes diversity in television news against the backdrop of theories about the public sphere and the media. Her research interests are journalism and the public sphere, environmental governance, and environmental communication. Several of her recent research projects and publications focus on public participation in environmental discourse.

Callum McGregor is a PhD candidate in education at the University of Edinburgh, supervised by Jim Crowther. He holds an MSc in community education and an MSc in educational research. His current research focuses on the cultural politics of contemporary environmentalism in the United Kingdom through studying the public discourse of various activist milieus. He continues to work part time as a community educator.

Robert Næss is an associate professor in the Department of Interdisciplinary Studies of Culture at the Norwegian University of Science and Technology (NTNU). His main interests are science and technology studies, and he has published articles in national (Norwegian) and international journals on climate adaptation, public understanding of climate change, and development of new renewable energy technologies.

Israel Parker is a postdoctoral researcher in the Department of Wildlife and Fisheries Sciences at Texas A&M University. His primary research focuses on wildlife ecology and management, particularly with over-abundant or nuisance populations and endangered or at-risk species, with side ventures into public participation and social movements pertaining to wildlife conservation and climate change.

Tarla Rai Peterson is the Boone and Crocket Professor of Wildlife and Conservation Policy at Texas A&M University, a guest professor of environmental communication at the Swedish University of Agricultural Sciences, and an adjunct professor of communication at the University of Utah. She studies intersections between communication and democracy with the goal of facilitating community-based conservation that contributes to sustainable public policy. Her current research centers on climate change communication, especially as it relates to energy policy. She maintains an active theory-to-practice program that includes design and evaluation of best practices for facilitating public participation in science and technology issues related to environmental and energy policy. She has published articles and book chapters in numerous academic and trade outlets. Her books include *Sharing the Earth: The Rhetoric of Sustainable Development*, *Green Talk in the White House: The Rhetorical Presidency Encounters Ecology*, and *Social Movement to Address Climate Change: Local Steps for Global Action*.

Andrea Polli is currently an associate professor of art and ecology with appointments in the College of Fine Arts and the School of Engineering at the University of New Mexico. She holds the Mesa Del Sol Endowed Chair of Digital Media and directs the Social Media Workgroup, a lab at the University's Center for Advanced Research Computing. Polli's artwork with science, technology, and media has been presented worldwide and has been recognized by numerous grants, residencies, and awards, including a NYFA Artist's Fellowship, a Fulbright Specialist Award, and the UNESCO Digital Arts Award. Her most recent book is *Far Field: Digital Culture, Climate Change, and the Poles* (with J. D. Marsching), available from Intellect Press.

Marianne Ryghaug holds a PhD in political science from the Norwegian University of Science and Technology (NTNU). She is professor of science and technology studies in the Department of Interdisciplinary Studies of Culture, NTNU, and deputy director of the Centre for Renewable Energy Studies (CenSES). She has conducted research in energy and climate since 1999 and has published widely on these topics in top international journals, such as *Public Understanding of Science* and *International Studies of Philosophy of Science*. Her research interests are energy and climate-related science communication, policy, and implementation and the use of new sustainable energy technologies.

Eurig Scandrett is a lecturer in sociology at Queen Margaret University, Edinburgh. After an initial career as a research scientist in plant ecology, he spent 15 years in adult education, including as head of community action at Friends of the Earth Scotland. His research interests focus on knowledge generation and distribution in environmental-justice movements, and he has published on lifelong education, environmental justice, and the political ecology of knowledge. He edited *Bhopal Survivors Speak: Emergent Voices from a People's Movement* and (with Budd Hall, Darlene Clover, and Jim Crowther) *Learning and Education for a Better World: The Role of Social Movements*. He is an active campaigner for environmental and social justice.

Sarah Schweizer is a program associate at the International START secretariat in Washington, DC. She coordinates and manages the implementation of activities under START's risk management, vulnerability, and adaptation portfolio. Her research activities and practice focus broadly on capacity building and human-environment interactions and, more specifically, on climate change adaptation, participatory research, social learning, and communication. Her work has appeared, among other places, in *Science Communication* and *Human Dimensions of Wildlife: An International Journal*. Sarah holds a master of science in the human dimensions of natural resources and a bachelor of science in environmental communication from Colorado State University.

Jessica L. Thompson (PhD, University of Utah) is an assistant professor in the Communication and Performance Studies Department at Northern

Michigan University. She has published on climate change communication and public lands in numerous journals, including *Science Communication, Society & Natural Resources, Environmental Modeling & Software,* and *Human Dimensions of Wildlife.*

Lightning Source UK Ltd.
Milton Keynes UK
UKOW040625211212

203985UK00002B/6/P